An Introduction to Machine Learning

Gopinath Rebala • Ajay Ravi • Sanjay Churiwala

An Introduction to Machine Learning

Springer

Gopinath Rebala
OpsMx Inc
San Ramon, CA, USA

Ajay Ravi
San Jose, CA, USA

Sanjay Churiwala
Hyderabad, Telangana, India

ISBN 978-3-030-15728-9   ISBN 978-3-030-15729-6 (eBook)
https://doi.org/10.1007/978-3-030-15729-6

Library of Congress Control Number: 2019936169

This Springer imprint is published by the registered company Springer Nature Switzerland AG
The registered company address is: Gewerbestrasse 11, 6330 Cham, Switzerland

Preface

Machine learning (ML) is the new age electricity. When electricity was discovered, nobody had anticipated the far-reaching effect it would eventually have. Machine learning has the same potential. Many people also consider machine learning (or artificial intelligence) as Industrial Revolution 4.0. There is often an apprehension that ML will cause massive loss of jobs. Our belief is that ML will not cause loss of jobs. Just like advances in electricity and computerization brought in more jobs due to increased productivity, ML has potential to create different kinds of jobs. What is needed is to be future-ready for the shift in nature of jobs.

We often hear people talking about machine learning. Some of them talk in big mumbo jumbo words. On the other hand, some people claim that it is just an extension of statistics, probably trying to assure themselves that it is nothing new. To us, ML is as much an extension of statistics as multiplication is just a repeated addition. "Multiplication is a repeated addition" may be true only for the introduction of multiplication in second grade. However, multiplication is then studied as a topic by itself, which allows us to apply multiplication to fractions, decimal numbers, vectors and matrices, concept of indices, etc. Similarly, ML is an extension of statistics if you consider only the introductory aspects of ML.

Many people believe ML only applies to complex activities requiring human-level intelligence such as automated driving and playing video/board games better than humans; however, ML has many other much simpler applications. So, ML is not just useful for the elite few. It is for anyone who has interest and willingness to apply it.

This introductory book demystifies ML. It assumes no prior knowledge of ML. We have given more importance to conceptual understanding of the underlying concepts and the algorithms involved in ML and gradual evolution from basics. We have intentionally kept the book to be software language/package and hardware implementation agnostic. Once you understand the concepts, you can implement your algorithms in the language and hardware of your choice.

All three of us have known each other since 1989, when we joined IIT Kharagpur in India for our undergraduate studies in Electronics and Electrical Communications

Engineering. Since then, we have been in close touch, worked with same employer, shared apartment and car, worked with intensely competing employers, and went through various other changes in our respective lives. We have seen each other evolve – both professionally and on personal front.

We learnt ML out of our own interest, not necessarily combined with our job responsibilities. During the course of our social conversations, we realized that we were either working in this field or passionate about it and felt like creating a book to explain the concept in a relatively simplified form. And, when we approached Springer with the idea, they were kind enough to help us share our learnings with the readers by publishing this book.

Reading the Book

Machine learning is a vast area, with application in many diverse fields. For practical reasons, we had to organize the chapters in a sequential manner. However, it is not necessary to read the whole book in the same sequence. Depending on your areas of interest, you can decide to skip certain chapters.

Chapter 1 covers the basics of machine learning. It also revises some of the underlying mathematical concepts. Most probably, you have already studied these concepts during your high school or undergraduate level. Depending on how long ago that was, you might want to revise these portions enough to regain familiarity and comfort. Subsequent chapters will make use of these concepts. Having a good comfort will allow you to easily follow the mathematical portions of the discussions. Chapter 2 provides a basic framework of learning models.

Chapters 3 and 4 start with the actual machine learning. Though, these are relatively simple concepts, most machine learning topics build on these concepts. At this stage, it should be possible for you to start creating simple applications related to machine learning, though complex and very cool applications like autonomous driving, playing games, etc. are still far away!

Chapters 5 and 6 are related to classification and clustering. This allows you to create some additional applications or to refine your existing applications based on specialized classification problems/needs. You can skip these chapters if you are not interested in clustering- or classification-type applications.

Chapters 7 and 8 are about further refining your algorithm, data, and decision process for the same applications that you could create based on your learnings of chapters so far.

Chapter 9 covers neural network. This is the beginning of your ability to create more complex relations for machine learning applications. Most of the very high-end applications use neural network as their basis. Chapters 10 and 11 provide some popular application domains of neural network.

By now, you are into very heavy-duty data-crunching. So, Chap. 12 shows ways to reduce datasets. Chapters 13 through 17 continue to show additional domains of machine learning applications, some of which use neural network as their basis.

Depending on your specific interest, you can decide to skip chapters covering applications of no interest to you. Once you have a thorough understanding of all these chapters, it should be possible for you to design whatever machine learning system you choose. The only thing standing in your way would be your willingness to try, or access to the required set of data and compute resources!

Chapter 18 provides additional tips for creating complete application systems, which make use of more than one component of machine learning. So far, the book was independent of your choice of hardware and software. In this chapter, we introduce you to the choices available to you.

Acknowledgments

Though the book lists the three of us as authors, like most good things, this work also has direct and indirect contribution from many people. We are indebted to many of our friends and colleagues, with whom we have discussed aspects of machine learning, which helped refine our own understanding of this field.

We want to acknowledge the time and expertise of Padmini Gopalakrishnan, Ph. D.; Jayanta Choudhury, Ph.D.; Shreesha Mahishi; Tushar Maniktala; Murali Krishnan; and Sasidhar Koppolu who reviewed portions of the book to help improve both accuracy and clarity of presentation.

We also want to thank our immediate family members who were patient enough for over a year, as we spent time on this work after coming home from our day job and over the weekends. They were not only patient enough but provided us with the encouragement to continue. Our extended family members and close friends have also been highly encouraging. Their feeling and expression of pride was a huge morale booster to us.

Thanks are also due to Charles Glaser of Springer for agreeing to bring this book to life. Our management chain at office was also highly encouraging of this work by us.

Hyderabad, Telangana, India — Sanjay Churiwala
San Jose, CA, USA — Ajay Ravi
San Ramon, CA, USA — Gopinath Rebala
April 2019

Contents

List of Figures

List of Tables

Chapter 1
Machine Learning Definition and Basics

Machine learning algorithms have shown great promise in providing solutions to complex problems. Some of the applications we use every day from searching the Internet to speech recognition are examples of tremendous strides made in realizing the promise of machine learning.

In this chapter, we discuss the definition of *machine learning* (ML) and its relation to *artificial intelligence* (AI). In addition, we will refresh basic mathematical concepts and algorithms that form the basis of ML.

It is important to grasp the concepts mentioned in this chapter as they are vital to understanding machine learning algorithms discussed in the rest of the book. Feel free to skip sections you already are very comfortable with. On the other hand, if some concept is not familiar at all, refer to a relevant mathematics book (typically, high school level) to get a good grasp on the topic. Most of the mathematical descriptions in this book are kept informal to keep the focus on understanding the concepts. The aim is to refresh, rather than to provide a primer on the underlying mathematical topics.

1.1 Introduction

Machine learning (ML) is a field of computer science that studies algorithms and techniques for automating solutions to complex problems that are hard to program using conventional programing methods. The conventional programming method consists of two distinct steps. Given a *specification* for the program (i.e., *what* the program is supposed to do and not *how*), first step is to create a detailed design for the program, i.e., a fixed set of steps or rules for solving the problem. Second step is to implement the detailed design as a program in a computer language.

This approach can be challenging for many real-world problems for which creating a detailed design can be quite hard despite a clear specification. One such example is detecting handwritten characters in an image. Here you are given a

G. Rebala et al., *An Introduction to Machine Learning*,
https://doi.org/10.1007/978-3-030-15729-6_1

dataset consisting of a large number of images of handwritten characters. Additionally the *data points* (i.e., images) in the dataset are *labelled*, i.e., each image is tagged or marked with the character it contains. This *labelled* dataset is essentially a set of examples describing how the program should behave. The objective is to come up with a program that can recognize the characters in any (new) image and not just the ones in the dataset. With a conventional method, you would first study the examples in dataset, trying to understand how the images correspond to characters, and then come up with a general set of rules to detect characters in any image. Creating such a set of rules can be quite challenging given the large variation in handwritten characters.

The general approach of crafting rules to solve complex problems was tried quite unsuccessfully in the 1980s, through *expert systems*. Moreover, there are so many of these challenging problems that solving them all individually in a conventional manner is simply impractical.

ML algorithms can solve many of these hard problems in a generic way. These algorithms don't require an explicit detailed design. Instead they essentially learn the detailed design from a set of *labelled* data (i.e., set of examples illustrating the program's behavior). In other words, they learn from data. The larger the dataset, the more accurate they become. The goal of an ML algorithm is to learn a *model* or a set of rules from a *labelled* dataset so that it can correctly predict the labels of *data points* (e.g., images) not in the dataset.

ML algorithms solve the problems in an indirect way, by first generating a *model* based on processing the dataset and then predicting the label of a new input data point by executing that *model.* This approach is known as *supervised machine learning*.

ML algorithms tend to be more accurate than human-created rules since ML algorithms will consider all data points in a dataset without any human bias due to prior knowledge.

While ML algorithms work well, it is still not very clear as to exactly how a problem is being solved. This is especially true for ML algorithms based on *neural networks*. This is called the *model interpretability* problem. In some problem domains, the *interpretability* of an ML algorithm is also equally important. For example, during a war, a military general would like to understand how the algorithm arrived at a decision before trusting the algorithm's decision. *Model interpretability* is an area of active research and is not discussed in this book.

1.1.1 Resurgence of ML

ML research made slow and steady progress on solving complex problems until the mid-2000s, after which the progress in the field accelerated drastically. The reasons for this dramatic progress include:

- Availability of large amount of data due to the Internet, such as large datasets of images
- Availability of large amounts of compute power, supported by large memory and storage space
- Improved algorithms that are optimized for large datasets

Some of the most prominent advancements in ML include the following:

- Speech recognition – the ability to recognize speech and convert it to text. This was a hard problem to solve until recent progress in ML.
- Language translation – understanding and forming language constructs without formal training in grammar, just like a child learning its mother tongue. Recent progress in ML resulted in a huge increase in the accuracy of translation compared to earlier methods.
- Driverless vehicles – the ability to navigate the vehicle without human intervention. This requires several capabilities that are enabled by recent progress in ML, including *perception* (identifying and tracking stationary and moving objects).

Compared to the era of the 1990s, where AI (artificial intelligence) was in the domain of the brightest engineers and scientists, ML/AI is now more accessible to general population, thereby raising the overall interest.

1.1.2 Relation with Artificial Intelligence (AI)

Artificial intelligence (AI) is a much broader field of study than machine learning (ML). AI is all about making machines intelligent using multiple approaches, whereas ML is essentially about one approach – making machines that can learn to perform tasks. An example of an AI approach that's not based on learning is developing *expert systems*.

Although intelligence is hard to define, it is clear that ML is a subfield of AI. A vast majority of people believe that AI and ML are the same due to a strong belief that ML is the only viable approach to achieving AI's goals.

The field of AI has been around for over six decades, and it went through a few hype and bust cycles during this period. During the hype period, AI research would show a lot of promise; however, when the promise didn't materialize, it was often followed by a bust cycle. During the bust cycles, the interest in AI waned (less funding for research, etc.). The last bust cycle was in the late 1980s and early 1990s. The AI bust cycles are popularly known as *AI Winters*.

The current excitement in ML and AI is different from previous AI hype cycles, because unlike previous hype cycles, ML techniques are being deployed to solve real-world problems in daily lives.

1.1.3 Machine Learning Problems

Machine learning can provide insights into structures and patterns within large datasets. It is also used to create models by learning from existing datasets to predict or forecast outcomes or behavior.

Some examples of machine learning problems include:

- *Classification*: Classify something into one or more categories (classes). For example, identifying if an email is a spam or not. Another example is identifying an image of an animal as a cat, dog, or some other animal.
- *Clustering*: Divide up a large set of data points into a few clusters, such that points within a cluster have some properties in common. Unlike classification, the number of clusters may not be known beforehand.
- *Prediction*: Based on historical data, build models and use them to forecast a future value. For example, demand for a particular product around holiday season

We first need to understand the basic building blocks behind the solutions of ML problems, before we can understand and build complex and useful applications. To this end, the next few sections cover the basic mathematics needed for ML.

1.2 Matrices

A *matrix* (plural *matrices*) refers to a set of numbers (or objects) arranged in rows and columns. The advantage of matrices is that it can often simplify representing larger amount of data or relationship. Matrices exhibit certain properties and these properties are well studied. That means, once we have a problem specified in the form of matrices, we can exploit matrix properties to solve the problem. Plus, there are many mathematical packages available for matrices that readily provide implementations for matrix manipulations.

For example, Fig. 1.1 shows matrix A with four rows and three columns. This matrix is represented as $A_{4\times3}$. Matrix A is said to be of dimension 4×3 (called *4 cross 3* or *4 by 3*). And, A_{12} represents the value at the intersection of Row 1, Column 2 of matrix A. It is also represented as $A[1,2]$. In Fig. 1.1, it would mean value 5.

Fig. 1.1 Matrix A with four rows and three columns

$$A = \begin{bmatrix} 7 & 5 & 9 \\ 2 & 8 & 4 \\ 6 & 13 & 1 \\ 5 & 11 & 10 \end{bmatrix}$$

1.2.1 Vector and Tensors

A *vector* is an ordered set of *N* elements, also called *n*-dimensional vector. It can be represented as a matrix with just a single column.

Tensors are a generalization of vectors in higher dimensions. A *tensor* of two dimensions is a matrix. *Tensors* with higher dimensions are represented by multidimensional arrays (which are supported in all ML programming libraries).

Tensors used in ML (distinct from *tensors* used in mathematics and physics) are synonymous with multidimensional arrays in programming.

Section 1.5 talks about the significance of vectors and their usage in linear algebra.

1.2.2 Matrix Addition (or Subtraction)

Two matrices *A* and *B* can be added to (or, subtracted from) each other, if they have the exact same dimensions. The resultant matrix *C* also has the same dimension. Each element of *C* is obtained by adding (or, subtracting) the elements at the same corresponding position of *A* and *B*.

For example, $A_{4\times3} + B_{4\times3} = C_{4\times3}$

```
For each I (1 to 4)
   For each J (1 to 3)
      C_IJ = A_IJ + B_IJ
   End for
End for
```

1.2.3 Matrix Transpose

Matrix transpose refers to swapping of rows and columns of a matrix. A^{T} or *Tr*(*A*) denotes transpose of matrix *A*. Figure 1.2 shows a matrix *A* and its transpose A^{T}. Satisfy yourself for few values that $A[I, J] = A^{\mathrm{T}}[J, I]$.

$$A = \begin{bmatrix} 1 & 2 & 3 \\ 4 & 5 & 6 \\ 7 & 8 & 9 \\ 10 & 11 & 12 \end{bmatrix} \qquad A^{\mathrm{T}} = \begin{bmatrix} 1 & 4 & 7 & 10 \\ 2 & 5 & 8 & 11 \\ 3 & 6 & 9 & 12 \end{bmatrix}$$

Fig. 1.2 Matrices *A* and A^{T}

1.2.4 Matrix Multiplication

Usually, multiplication operation for matrix is represented by *X*; however, in this book, we will often represent it as * or sometimes without any explicit operator (similar to algebraic expression style) so that it is not mistaken for matrix named *X*.

1.2.4.1 Multiplying with a Scalar

A *scalar* is a unit value that usually represents magnitude. The result of multiplying a matrix *M* with a *scalar* is a matrix of the same dimension as the matrix *M*. Each element of the resulting matrix is obtained by multiplying the *scalar* with each element of matrix *M*.

For example, $s * A_{4\times3} = C_{4\times3}$, where *s* is a *scalar*

```
For each I (1 to 4)
   For each J (1 to 3)
      C_IJ = s*A_IJ
   End for
End for
```

1.2.4.2 Multiplying with Another Matrix

A matrix *A* with dimensions $m \times n$ can only be multiplied with another matrix *B*, if *B* has a dimension $n \times p$, i.e., the number of columns of *A* should equal the number of rows of *B*. The resulting matrix *C* has the dimension $m \times p$, i.e., the same number of rows as *A* and the same number of columns as *B*.

For example, $A_{4\times3} \times B_{3\times5} = C_{4\times5}$

```
For each I (1 to 4)
   For each J (1 to 5)
      C_IJ = 0
      For each K (1 to 3)
         C_IJ = C_IJ + A_IK * B_KJ
   End for
End for
```

Since matrix multiplication takes some extra effort to comprehend, look at it in another way. For any element C_{IJ}, look at Row *I* of matrix *A*. You should get three elements. Now, consider Column *J* of matrix *B*. Again, you will get three elements. Multiply each of the three elements obtained from *A* by the corresponding three elements obtained from *B*. You should be getting three results. Add these three numbers. This is element C_{IJ}.

Consider the example matrices shown in Fig. 1.3.

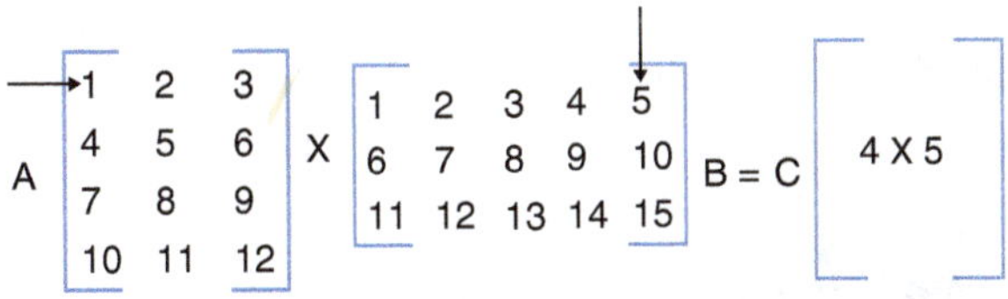

Fig. 1.3 Matrices *A* and *B* multiplied

$$C_{15} = A_{11} * B_{15} + A_{12} * B_{25} + A_{13} * B_{35} \tag{1.1a}$$

$$C_{15} = 1 * 5 + 2 * 10 + 3 * 15 = 70 \tag{1.1b}$$

Clearly, matrix multiplication is not commutative, i.e., $A * B$ is not the same as $B * A$. Further, depending on dimensions of A and B, sometimes, $A * B$ or $B * A$ or neither of them may not even be possible.

1.2.4.3 Multiplying with a Vector

You can consider multiplication of a *matrix* with a *vector* as a special case of Sect. 1.2.4.2, where the matrix B has only one column. A matrix A with dimensions $m \times n$ can be multiplied with vector B, if B has n rows, i.e., the dimensions of B is $n \times 1$. The resulting *matrix* C has the dimension $m \times 1$ clearly, the resultant matrix C will also be a *vector*.

1.2.5 Identity Matrix

Identity matrix is a special type of matrix, which has the following important properties:

- It is a square matrix (i.e., has the same number of rows as the number of columns).
- All its diagonal elements are 1.
- All its non-diagonal elements are 0.
- It is often represented as I.
- When multiplied by another matrix, you can work out that the resultant matrix will be the same value as the original matrix, i.e., $A * I = I * A = A$.

1.2.6 Matrix Inversion

Matrix *inversion* of A results in another matrix, often denoted by A^{-1}. This matrix has the property that when multiplied by A, the resultant matrix is I. So,

$A * A^{-1} = I = A^{-1} * A$. Only square matrices can have inverse, though not all square matrices will have an inverse.

1.2.7 *Solving Equations Using Matrices*

We often try to express ML problems in matrix form both for clarity and to use matrix properties to solve the problems. The exact problem formulations and their solutions will be explained in subsequent chapters. However, to get an idea, consider the following problem representation:

A set of equations can be expressed in two forms. One of the forms is the equations themselves:

$$3x + 4y + 5z = 26 \tag{1.2a}$$

$$2x + 5y + 7z = 33 \tag{1.2b}$$

$$x + y + z = 6 \tag{1.2c}$$

The second form is in terms of matrix, $A * X = B$, where A, X, and B are as shown in Fig. 1.4.

Once expressed in matrix form, $A * X = B$, the value of X can be obtained as:

$$A^{-1} * A * X = A^{-1} * B \tag{1.3a}$$

$$\text{Or} \quad X = A^{-1} * B \tag{1.3b}$$

Thus, the *simultaneous equations* could be solved through *matrix inversion* and multiplication!

A few other advantages of using matrix-based problem formulations are as follows:

- Most software packages for mathematical analysis have specific functions and utilities for matrices.
- Many ML problems can exploit parallel computing technique available for matrices.

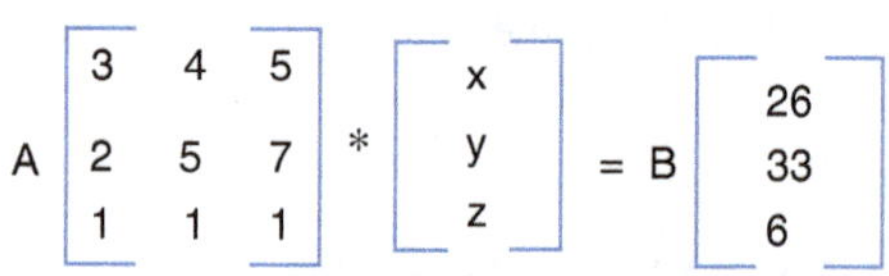

Fig. 1.4 Matrix representation for simultaneous equations

1.3 Numerical Methods

Numerical methods are algorithms for computing approximate values of mathematical expressions, typically using simpler computations. They can be iterative methods where each iteration improves the accuracy of the value being computed.

Let us consider a simple example of an iterative method. Suppose you want to find the positive square root of 25, without actually knowing how to compute square root.

We first guess the answer as 1. We see that its square is 1, which is significantly less than 25. Therefore the square root should be greater than 1. Then we increase it to 10 in next iteration.

Now, the square of 10 is 100, which is way higher than 25. That means, the square root should be less than 10. So, we need to choose a number below 10 – but above 1. We can go through this iterative process until we reach a value whose square is 25 (or, close enough for your purpose), and that would be the answer. The actual algorithm will be described a few paragraphs later.

Numerical methods are fundamental to ML problems, since we often can't compute the exact solution. Most of the time, there is no *closed-form solution* (i.e., a mathematical formula) for a function to be computed, and numerical methods have to be used to compute an approximate solution. Even if a closed-form solution exists, computing it may be so tedious that we may resort to numerical methods to get an approximate solution.

In the coming chapters, we will state the exact problem and the optimal solution based on numerical methods, applicable for the specific problem formulation. This section is just to let you get a feel of numerical methods.

Let us work through the square root problem now. You will see that the following solution converges fast. We have chosen to compute square root of 25. You can try picking any other number (Table 1.1).

Algorithm to iteratively compute square root using numerical analysis:

Step 1. Pick some value C for the candidate answer (i.e., possible value for square root). Say $C = 1$.

Step 2: Compute quotient $Q(25/C)$.

Table 1.1 Example solution through numerical analysis

Iteration	Candidate value (divisor)	Quotient (=25/divisor)	Next candidate value (average of divisor and quotient)
1	1	25	13
2	13	1.92	7.46
3	7.46	3.35	5.4
4	5.41	4.62	5.02
5	5.02	4.98	Converged!!

Step 3: Compare quotient Q with divisor C. If their values are close, i.e., differ by less than 0.5, then their average is the square root, and we are done! If not, go to Step 4.

Step 4: Take the average of the divisor C and the quotient Q, and use this as the next candidate C, i.e., $C = (C + Q)/2$. Go to Step 2.

1.4 Probability and Statistics

You will see dependence on probability concepts to formulate ML problems in various chapters. We will avoid intense mathematical derivations at many places, when they are not directly relevant to ML optimizations.

A number of things in this world appear to be random or unpredictable, e.g., tossing a coin. Many other phenomena in real world have an element of *randomness* (also known as *stochastic phenomena*) such as stock market index value, temperature at a given time, the number of people in a city at any time, etc. Probability indicates how likely a specific outcome is, of an observation or a measurement. For example, if you pick a ball at random from a basket containing 2 green balls and 1 blue ball, the probability of selecting a green ball is 2/3, and the probability of selecting a blue ball is 1/3.

Since the outcomes are all mutually exclusive, and since one of them will certainly occur, the sum of their probabilities is 1.

The *probability distribution* shows the probability values for each of the possible outcomes. In many real-world situations, it is easier to specify the distribution using some parameters, rather than probability for each possible outcome. For example, the distribution of weights of people in a large population can be represented through *mean* and *variance* of the weights rather than probabilities of specific weights. The numeric values represented could be *discrete* (e.g., number of adults in a household) or *continuous* (e.g., the weight of an individual).

When the numeric values are continuous, you will only consider the probability for a range of values rather than a specific value. The probability distribution for continuous variables is described by a *probability density function* (*pdf*). The *pdf* is defined in such a way that the probability over the whole range of values for the random variable is 1. You can consider the values of *pdf* as indicative of actual probabilities. Normal distribution (explained in Sect. 1.4.5) is an example of such probability distribution function.

The easiest distribution is the *uniform distribution* where all events are equally likely.

1.4.1 Sampling the Distribution

Sometimes, you do not know the inherent characteristics of the underlying elements to be able to compute the probability, e.g., if a coin is defective or if you don't know the contents of the basket – in terms of how many balls are of which color. In that case, you draw many samples, and based on that you estimate the original distribution. For example, you randomly draw balls from the basket, many times. Depending on how many times you obtained a green ball, and how many times blue ball, you will be able to get an estimate of the composition of the basket – in terms of colorwise count of the balls. This estimate may not necessarily be exact. However, the more samples you consider, the closer you will keep getting to the correct distribution.

1.4.2 Random Variables

In most random phenomena, there is a numeric quantity associated with a phenomenon that varies randomly, such as weight of a person or the value of stock price. This numeric quantity represents the value of a variable, say X. However, sometimes the event is not naturally a number.

In the world of mathematics and statistics, it is always easier to deal with numbers. So, you can decide to assign a number to the events. For example, drawing a green ball could be associated with X being 3, and drawing a blue ball could be associated with X being 9. So, instead of saying *P*(*green ball*) = 2/3, you would say $P(X = 3)$ is 2/3. The choice of variable value is usually made with some deliberation to solve the specific problem at hand.

1.4.3 Expectation

One of the important numeric values associated with a probability distribution is the *expectation*. It is essentially the average of an infinite number of *samples* from the probability distribution. The average of a large number of *samples* will approximate the *expectation*.

Suppose X takes on values $x_1, x_2, \ldots, x_N$, with corresponding probabilities p_1, $p_2, \ldots, p_N$. Then expectation $E(X)$ is the weighted average given by $p_1 * x_1 + p_2 * x_2 + \cdots + p_N * x_N$. Consider the same example of green and blue balls, where the random variable is assigned a value of 3 (for green balls) with probability 2/3 and value of 9 (for blue balls) with probability 1/3. The *expectation* is 5. Therefore the average over a large number of samples of balls drawn will be close to 5.

1.4.4 Conditional Probability and Distribution

Conditional probability of an event A given an event B is written as $P(A \mid B)$. Bayes' theorem states that the value of this conditional probability is $\frac{P(A \text{ and } B)}{P(B)}$. The *conditional probability distribution* of A, given a specific value b of B, is written as $P(A \mid B = b)$.

1.4.5 Maximum Likelihood

Most problems in machine learning are based on finding maximum likelihood or in other words those having highest probability. To get a feel of what we mean by maximum likelihood, let us consider a random variable that follows a *normal distribution*. The pdf of a *normal distribution* has the shape as shown in Fig. 1.5.

The curve is characterized by two parameters:

- Mean: this is the value at which the curve shows the highest value.
- Standard deviation: this represents the spread of the curve. Higher standard deviation means the curve is flatter and has a higher spread.

Given a normal distribution curve (i.e., mean and standard deviation specified), it is relatively simple to find the probability of various values. In ML, the problems are the other way. Given a large number of sampled values for a random variable, identify the normal distribution which is most likely to generate that set of values.

Consider the points shown in Fig. 1.6, obtained through sampling of a random variable.

ML aims to identify which distribution is most likely to generate the data as shown in Fig. 1.6. For example, Fig. 1.7 shows three different curves. The curve (*b*) is most likely to generate the data as observed in Fig. 1.6.

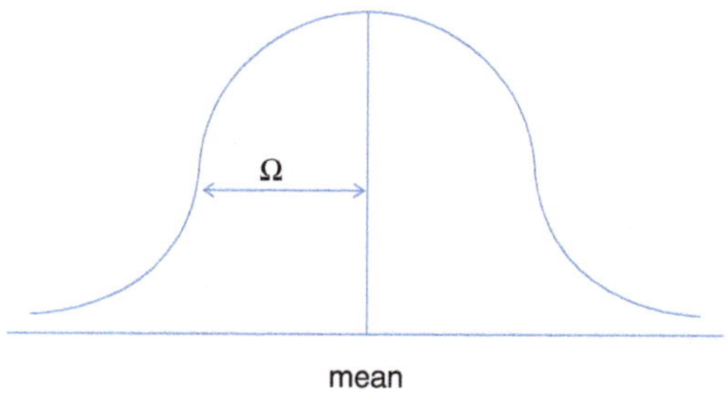

Fig. 1.5 Normal distribution curve

Fig. 1.6 Points observed/obtained through prior data

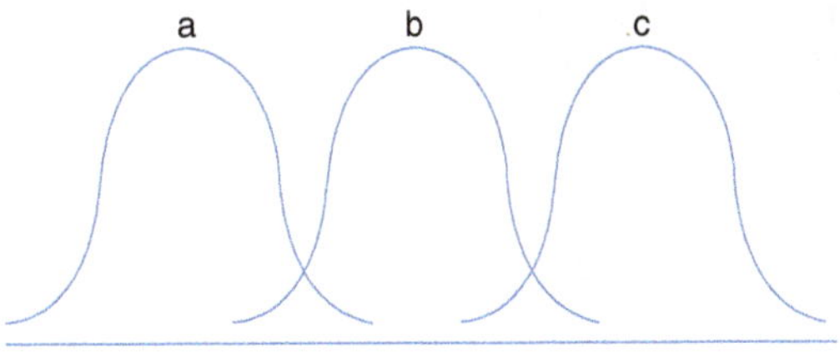

Fig. 1.7 Likely curves that can generate the data shown in Fig. 1.6

There are two methods to determine the most likely normal distribution for a given set of observations:

- Directly determine the normal distribution's parameters *mean* and *standard deviation* through knowledge of statistics.
- Assume some parameters of the normal distribution, and use numerical analysis methods to gradually modify the parameters towards the desired normal curve.

The data for ML problems can follow various probability distributions. Knowing the distribution of data lets us solve for the most likely (maximum likelihood) parameters of that distribution. And, depending on the exact problem to solve, we will use combinations of mathematical formulas and numerical analysis methods.

1.5 Linear Algebra

Usually, we can represent any two-dimensional relationship through a graph. For example, a point (2, 3) represents a point on a graph with two axes (*X* and *Y*), where the *X*-value is 2 and *Y*-value is 3. Even a three-dimensional relationship can be mentally visualized and can also be represented on a graph, for example, (2, 3, 4) to represent the values on three axes (*X*, *Y* and *Z*). However, it is quite difficult to visualize a multidimensional relationship. Even though a coordinate system with more than three axes is difficult to visualize, we can still represent a point on a five-axis system through numbers only. For example (2, 4, 7, 6, 1) represents a point with value 2 on first axis, 4 on second axis, 7 on third axis, and so on.

Another mechanism of representing such points on a multidimensional space is through a *vector* (matrix with just one column) as discussed in Sect. 1.2.1. So, the above point would be represented as shown in Fig. 1.8.

In machine learning, we deal with a lot of *high-dimensional* data, i.e., each data is a vector with a large dimension. For example, if you want to predict the closing prices of today's stock index, you might be considering many *factors* that affect today's index, such as last 200 days of closing price, today's opening price, trade volume of several days, etc. Let us say, you consider 10 *factors* for prediction, then each of these 10 *factors* is represented by a number, and each *set* of these numbers can be represented by a single 10-*dimensional vector*. The ability to express data points as multidimensional vector allows us to formulate the ML problems

Fig. 1.8 Multidimensional point represented as a vector

$$\begin{bmatrix} 2 \\ 4 \\ 7 \\ 6 \\ 1 \end{bmatrix}$$

compactly in terms of matrices so that we can exploit matrix operations for more efficient computation.

1.6 Differential Calculus

Differential calculus allows you to determine rate of change of a function at any point, with respect to one of the input variables. This translates to the *slope* of a curve (i.e., function drawn as curve) at any given point for the curve.

1.6.1 *Functions*

Concept of differential calculus needs you to be familiar with concepts of *functions* and how they represent relationship among a set of variables. Consider a function $F(x) = x^2$. Given the function, it is always possible to find a unique value of the function, corresponding to any given value of x. Sometimes, given a set of values, it might be possible to find the function – if you know the nature of the function and sufficient number of data representing the values. For example, if $F(x)$ is a linear function of x, it is possible to obtain the exact function, if you know two values of x and the corresponding values of $F(x)$. Sometimes, the exact nature of the function is not known. ML problems often involve approximating the exact functions (including its nature), given a set of observed (labelled) data. This is called *function approximation*.

1.6.2 *Slope*

Consider a curve for single input function with input x and output y, as shown in Fig. 1.9.

For point A, the slope is negative. That means, as you go toward the right of the curve from this point, the curve will be descending. For point C, the slope is positive. That means, as you go toward the right of the curve from this point, the curve will be ascending. And, at B, the slope is 0. The beauty of this concept of slope is that it

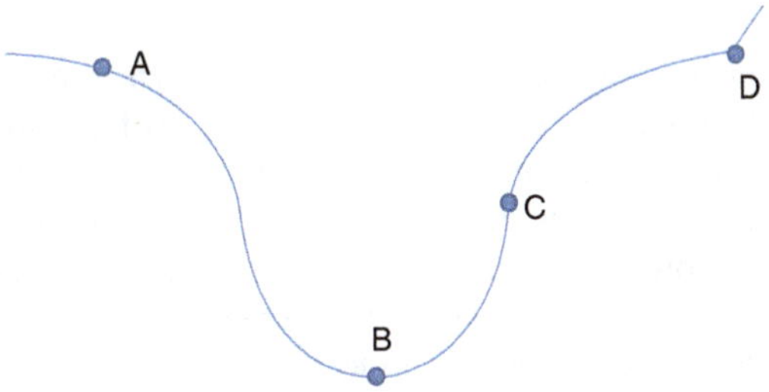

Fig. 1.9 Curve for showing the concept of slope

gives a very localized answer. It does not look at the whole curve. Just at that point! So, the slope just to the left of *B* will be negative and just to its right will be positive.

You can use the concept of slope to decide if you want to increase a value or decrease it – in order to follow a curve in a specific direction (up or down). Besides telling whether the curve will be going up or down as we move in any given direction, slope also gives an indication of how sharply does the curve change. For example, referring again to Fig. 1.9, while the slope is positive for both points *C* and *D*, the slope is higher at *D* – indicating that at point *D*, we will ride (or slide down) the curve much sharply, compared to point *C*.

And, the last important point of our interest in this domain is the significance of slope being 0. When the slope is 0, it indicates that the curve is at its minimal or maximal value at this location. The value may or may not be extreme for the whole curve. We only know that in localized context, the value is at its extreme. As you can see in Fig. 1.9, point *B* is at minimal value (considering the curve just around the point).

If *y* depends on variable *x*, it is possible to chart the relationship of *x* and *y* through a curve on two-dimensional axis. The differential of change in value of *y*, with small change in value of *x*, is represented by dy/dx at any given value of *x*. dy/dx represents the slope of the curve *y* (plotted against *x*), at exactly that point. In mathematical terms, this is called *derivative*. For many mathematical curves, it is possible to define the derivative. For example, if $y = x^3$, slope/derivative will be $3x^2$, and this would be valid for any value of *x* on this curve.

Let *z* be a variable that depends on both *x* and *y*, where *x* and *y* are independent of each other. This brings in the concept of slope of *z* with respect to *x* as independent from the slope of *z* with respect to *y*. These two slopes are represented as $\partial z/\partial x$ and $\partial z/\partial y$, also called *partial derivatives*.

1.7 Computer Architecture

Since machine learning involves a lot of computation on a lot of data, it will be good to know a little bit about computer architecture. This will allow you to optimize your algorithms to be faster.

Any data and instruction for a computer is stored in memory. Actual computation on the computer is done inside a CPU. The speed at which CPU executes instructions depends not only the speed of the CPU but also on the time it takes to access data from the memory. This is called the *memory access time*. The smaller the *memory access time*, the faster a CPU can execute instructions.

Now, computer memory is organized in hierarchies (or, layers). Memories that are very close to the CPU have the smallest *memory access time* (but they also tend to be most expensive, in terms of cost per memory bit), and the memories that are furthest tend to be the cheapest but have large memory access time.

Since ML works on large datasets, they have to be stored slightly far from the CPU (e.g., in onboard external memory or on SSD or hard drive). That means, the algorithms that do lot of memory accesses will tend to be much slower compared to another algorithm that minimizes memory accesses.

Consider an example. Say, you have a very large set of numbers, represented by $x1, x2, \ldots, x1000000$. You want to compute running average of 50 numbers. One way is to access $x1$ through $x50$ and compute the average. Next, retrieve $x2$ through $x51$, compute the average, and so on. As you can see, you are getting 50 numbers from the memory each time. Notice that we are repeatedly retrieving the same numbers (e.g., $x53$ is retrieved 53 times).

Another way is to compute the first average, using numbers $x1$ through $x50$. And, besides computing the average, also keep track of the sum of these 50 numbers. For the next average (from $x2$ to $x51$), you can just subtract $x1$ from this first sum and add $x51$, and this gives the next sum, from which you can compute the next average. In this algorithm, we only had two memory accesses for each subsequent average. Therefore this algorithm would be faster since it has significantly reduced total memory access time.

Chapter 18 gives a brief overview of some of the compute types and their suitability for ML-type problems, comparing them on large amount of data access and parallel compute capabilities.

1.8 Next Steps

In this chapter, we introduced various concepts. These act as good prerequisite to getting a good understanding on ML. A lot of these sections are topics of a complete book by themselves. It should be thus obvious that we have barely scratched the surface of those topics in this chapter. Fortunately, you don't need to be a master of these topics. A reasonable understanding will do.

The important thing though is to realize that ML is a multidisciplinary topic. This multidisciplinary nature, along with a generous usage of mathematical constructs, often makes the topic a bit daunting for many first-time learners of ML.

If you get a reasonable understanding of the topics explained here, you should be able to follow the following chapters. Before you go further, we would advise that you make yourself comfortable with the topics explained here. Ability to solve simple problems will be sufficient. In the subsequent chapters, multiple concepts will start interacting. Hence, having a comfort with each concept individually will make it easier to grasp those interactions.

Fortunately, as you venture into real-world ML problems, you will be able to make use of a lot of preexisting software packages. So, you can afford to not worry about a lot of implementation details of the computations. Still, solving a few problems yourself will get you a good feel of what is happening and how things are actually working.

- Solve a few simple problems related to each of matrix addition/subtraction, multiplication, inverse, and transpose, from any high school math book.
- Consider certain value of x and y. Say $x = 3$ and $y = 9$. Assume the relationship to be $y = mx$. See if you can use numerical method-based computation to find the correct value of m (3 in this case). Start with $m = 0$ as the initial guess.
- Solve a few simple problems related to finding the slope (derivative) at specific values, from any high school calculus book. For example, slope of the curve $y = x^3 + 3x^2$, at $x = 4$ and at $x = -2$. Also, draw the curve, and see how your answers correlate with the graph drawn.

Once you have solved these simple problems, you can go confidently to the next set of chapters.

Chapter 2
Learning Models

Machine learning algorithms derive their strength from their ability to learn from the available data. This chapter introduces you to the most prominent learning models. Subsequent chapters will dive deeper into the various algorithms or applications that depend on these models. The primary learning models that will be considered are:

- *Supervised Learning*
- *Unsupervised Learning*
- *Semi-supervised Learning*
- *Reinforcement Learning*

Before you go into the actual learning models, it is good to get introduced to the concept of *labelled* and *unlabelled* data. When you know the right answer to a question related to the data, then it is *labelled* data. For example, you have an image, and the question is what animal is in the image. If you know that the image represents a dog, then it is a labelled image.

When you don't know the right answer to the question, it is an *unlabelled* data. For example, you have a lot of images and you don't know the answer to which images belong to what category, then those are unlabelled data (images).

2.1 Supervised Learning

In *supervised learning* (SL), the machine is given a dataset (i.e., a set of data points), along with the right answers to a question corresponding to the data points. The learning algorithm is provided with a huge set of data points with answers, i.e., a *labelled* dataset. The algorithm has to learn the key characteristics within each data point in the dataset to determine the answer. So, next time a new data point is provided to the algorithm (machine), based on the key characteristics, the algorithm should be able to predict the outcome/right answer.

A few examples will help explain supervised learning:

G. Rebala et al., *An Introduction to Machine Learning*,
https://doi.org/10.1007/978-3-030-15729-6_2

Over a period of time, since its birth, a child is shown various objects and also told the right name for the object. The child gets to see his pet dog in many different positions, and when the child goes out, it gets to see many different dogs, in various positions. Gradually, the child starts recognizing dogs and is able to differentiate dogs from other nonliving objects and from other animals. During this process, once in a while, the child may make some incorrect observations, for example, identifying a wolf's picture also as a dog; and the parents will immediately teach that it is a wolf and not a dog. The child internally adjusts its understanding about what a dog is and gradually gets more and more accurate. Here, the child has been shown many pictures and real dogs, and it heard from parents, neighbors, teachers, elder siblings, etc. that they are all dogs. So, it's an example of supervised learning.

In the context of machine learning:

A machine (usually a computer program) is given thousands of pictures, and for each picture, the machine is given the right answer (the image is of a dog, cat, car, airplane, etc.), hence, giving a label to the picture. The machine then *learns* the characteristics within these pictures to be able to differentiate a dog from an airplane or from a cat.

A machine is given a lot of data corresponding to stock market. And for each data point, the machine is also told what happened to the stock (or the index) on that day, i.e., the label corresponding to the data point. The machine then *learns* the contribution of each of these data points into the final index/stock price. Now, when you present the machine with a new data point, the machine should be able to predict the stock investment choice – based on the respective contributions of each individual component of the dataset.

Supervised learning falls mostly in two categories; these are introduced briefly in this section, and you will start learning in detail from Chap. 3 onward.

2.1.1 Classification Problem

This refers to the ability to classify something into a distinct set of classes or categories. Recognition of an object into dog, cat, airplane, etc. is an example of identifying the class of the object.

Whether stock market is going to see a significant increase or decrease or no major change is another example of this category. Here the three classes are increase, decrease, and no change.

2.1.2 Regression Problem

Regression refers to the ability to predict values of a continuous variable, for example, a model to predict stock price on a daily or weekly basis. Predicting the number of transactions for an e-commerce site is another example.

2.2 Unsupervised Learning

In *unsupervised learning*, the machine is provided with a set of data and is not provided with any *right* answer. Given the huge amount of data, the machine may identify trends of similarity. The algorithm will identify clusters or groups of similar items or similarity of new item with existing group, etc.

When my daughter was a very small child, we had brought a toy for her. There was a cylindrical box with a cover on it. The cover had 3 holes in it (each with a different shape), and there were several (actually, 12) independent block pieces; such that each piece could go into only a specific hole. These pieces were of different colors. She had to put in all the block pieces into the box, placing each piece into the right hole. Over a period of time, she started figuring out, which piece would go into which hole. She still did not know the name of the shape; however, she knew that four blocks were similar in shape to a specific hole on the cover and other four blocks were similar to another specific hole. This is an example of unsupervised learning. It is unsupervised in the sense that once she knew a particular piece would go into a specific hole, she would be able to put all similar shapes into the same hole. Strictly speaking, there is an element of reinforcement learning also, since, as she tried various pieces into different holes, she would get a feedback, whether her efforts were successful or failing – and gradually learning from the experience.

In the context of machine learning:

The machine is given data on the buying patterns of all customers of a grocery store. Over a period of time, the machine is able to identify buying patterns, such as those who buy school bags also buy coloring pencils or those who buy beer also buy chips. The machine still does not know what these objects specifically are. However, it will be able to identify the correlation.

Another example you see in everyday life relates to the ready-made garment sizes. The world has billions of people, and each person is unique in terms of their body shape. It is not possible for ready-made garment manufacturers to create products for each unique body shape. Given various parameters of body shape, machines can identify clusters – a closely concentrated group where all the members are relatively similar to each other. And, so the manufacturer can make one size that would reasonably fit all members of this cluster or group.

Revisiting the example of identifying dogs, let us say, a machine is given thousands of pictures. The data is not labelled to identify which picture is a dog and which is a cat. After analyzing the characteristics of these pictures, the machine may identify similarity among certain pictures and consider them as one group with common characteristics. So the machine could identify all dogs being similar to each other but different from the group of cats. The machine would still not know which group is cat and which group is dog. It will just know that there are two distinct groups.

Chapter 6 on clustering provides detail on workings of unsupervised learning.

2.3 Semi-supervised Learning

Semi-supervised learning falls somewhere between supervised and unsupervised learning. Here, the machine is given a large dataset, in which only a few data points are labelled. The algorithm will use clustering techniques (unsupervised learning) to identify groups within the given dataset and use the few labelled data points within each group to provide labels to other data points in the same cluster/group.

One of the most prominent benefits of this technique is that you don't have to spend a lot of time and effort in labelling each data point, which can be a highly manual process.

Say, you want to create an algorithm to identify cats and dogs. For this, you need thousands of images of cats and dogs. And, for each image, you might need to tell the machine whether the image represents a cat or a dog. That means, you have to manually sift through each of these images and classify it as a cat or a dog.

Alternately, you could identify only few of the cat images and few of the dog images. Give all the images (mostly unlabelled) to the machine. The machine will find similar images among these, and all similar images will get the same label – derived from the few labelled images within this group. Hence, each image will now have a label that the machine can then use for its further learning. Clearly, semi-supervised learning is quite useful when you have a very high number of unlabelled data points, compared to the labelled data points. Semi-supervised learning can be quite challenging compared to supervised learning and is an active area of research.

2.4 Reinforcement Learning

Reinforcement learning is one of the most exciting areas of machine learning. Reinforcement learning is useful for situations involving:

- *Changing situations*: for example, driving, game of chess, backgammon, etc. Here, external situation (or, opponent's play) is continuously changing, and the response from the machine has to consider the changed environment.
- *Huge state space*: again, driving and multiplayer games are examples. Games like chess have almost infinite possible board configurations. It is not possible to play those games well using brute-force searching for moves, from an enumeration of all possible paths through the progress of the game. There are too many possible paths to even enumerate.

Hence, the learning has to result in the machine sensing the external environment and choose an action based on its own state and the external environment, with the aim of maximizing a specific predefined goal (e.g., staying within the lane for vehicular driving).

Given the huge amount of state space, the longer a machine is allowed to observe and learn, the better it is able to learn the longer-term impact of its decision, and hence, choices exercised by the machine start considering a longer-term benefit. For example, if you were to teach a machine to play the classic Atari breakout game using reinforcement learning techniques, within a very short span, the machine will be able to play the game reasonably well. However, when you train the machine for a few hours, it can learn to play at an expert level by creating a tunnel and try sending the ball up through the tunnel – so that the ball bounces off the bricks themselves, rather than coming down to be hit by the paddle.

Chapters 16 and 17 go cover details of reinforcement learning.

Chapter 3
Regressions

Regression analysis is a statistical and machine learning technique that allows you to build a model based on a *labelled* dataset (such as past data for stocks) that can be used to make predictions. This chapter starts getting into the actual aspects of machine learning. By the end of this chapter, you will be able to perform some simple predictions for new data, based on learning from the labelled dataset (from prior observations).

3.1 Introduction

Regression is an act of learning from existing data (can be past data) in trying to find relationships in the data. *Regression analysis* techniques help build relationships on various dimensions of the data to make predictions on the variable of interest. For example, based on past data, you can derive a model for a price of a real estate based on the interest rates, inflation, original purchase price, etc. There are two kinds of regression analysis:

- *Linear Regression*: This allows you to predict value of a continuous variable, for example, closing price of a specific company stock or of an index.
- *Logistic Regression*: This allows you to predict whether or not a specific outcome would be achieved. For example, whether or not a specific company stock will hit a certain price.

The rest of the chapter introduces you to concept of a *model* and methods of regression that are used in building the model.

G. Rebala et al., *An Introduction to Machine Learning*,
https://doi.org/10.1007/978-3-030-15729-6_3

3.2 The Model

Assume that the stock price at the end of a day is dependent on various factors. For simplicity, say these factors are closing price of last 10 days and the volume on each of these 10 days.

In this case, the model says, it is possible to find the price of a stock based on the formula in Eq. 3.1:

$$\begin{aligned}\text{Price} = &\text{ some constant} + a * \text{yesterday's closing price} \\ &+ b * \text{prior day's closing price} + c * 2 \text{ days back's closing price} \\ &+ \ldots + k * \text{yesterday's volume} + l * \text{prior day's volume} + \ldots\end{aligned} \tag{3.1}$$

If you can determine the values of *parameters* (*or coefficients*) a, b, c, etc., it would be possible to predict the value of tomorrow's closing price – after the market closes today.

Thus, the need is to find the correct values of a, b, c, etc., and these values can be computed based on data obtained from historical information.

For computing the model parameters, you will feed in historical data into the system as labelled *examples*. Each *labelled example* consists of an input/output pair. The prior 10 days' closing prices and the volumes are the input, and the closing price for that combination of data is the output. Finding the value of coefficients a, b, c, etc. that satisfies most of the historical data with acceptable error gives the model the ability to predict next day's stock price.

By now, it should be clear that *regressions* fall under *supervised learning*, since it requires labelled examples (such as past historical data).

3.3 Problem Formulation

Let y represent the variable that you want to predict. Continuing the stock price example from previous sections, it is the closing price of a stock/index. Let x_1, x_2, etc. be the variables that represent the factors (also called *features*) based on which you will predict y. For simplicity, the *features* are the closing price of the prior 10 days and the volume for the prior 10 days. Hence, you need 20 variables, $x_1, x_2, \ldots x_{20}$. in the model. Let $\theta_1, \theta_2, \ldots, \theta_{20}$ be the various *coefficients*, represented by a, b, c, etc. in Eq. 3.1. There is also a constant term represented by θ_0.

Thus, Eq. 3.1 can now be written in a more generic form, represented in Eq. 3.2:

$$y = \theta_0 + \theta_1 * x_1 + \theta_2 * x_2 + \theta_3 * x_3 + \ldots. \tag{3.2}$$

where

y = value to be predicted
x_i = variable for ith feature that impacts the value of y
θ_i = coefficient that determines the contribution of x_i to the value of y

We can also bring in a term x_0, such that its value is always 1. This allows you to use $\theta_0 * x_0$, instead of just θ_0. The advantage of introducing this new term is uniformity. Now, all the terms on the right-hand side of the Eq. 3.2 are very similar. This uniformity now allows you to write the equation in a further compact form presented in Eq. 3.3:

$$y = \sum_{i=0}^{n} (\theta_i * x_i) \tag{3.3}$$

where

n = number of features used in predicting y
x_i= *feature* at index i
θ_i= coefficient that determines the contribution of *feature* at x_i.

The way to interpret the $\sum$ symbol is sum of $\theta_i * x_i$, where i takes the values 0 through n. The following pseudo code shows the meaning of Eq. 3.3:

```
Initialize y = 0
For I = 0 to n
   y += θi * xi
End For
```

Equation 3.3 is also called a *hypothesis function* (or just *hypothesis*) and is represented by h_0. Now, the need is to find the appropriate values of *coefficients* θ. Once that is done, it would be possible to predict y.

Continuing with past example of stock prices, you need to determine a set of coefficients θ such that when the model is applied to historical data, the predicted values should be as close to the observed values as possible. The closer these values (predicted vs actual) are, the better is the accuracy of the model.

All data points in the dataset can be represented by Eq. 3.3. If there are m data points in the dataset, then there are m equations with different values of x_i and y with the same set of coefficients θ_i. These equations can be represented in matrix form $Y = X\theta$.

3.4 Linear Regression

You can use *Linear Regression* methods to predict values for a continuous variable, such as stock/index. *Linear Regression* assumes that the relationship between the variables can be expressed as a *linear function* (for a single input function, this is a straight line). It is possible to convert nonlinear relationships into a form that allows Linear Regression to be applied. This chapter focuses on solving problems with assumption of linear relationships.

Assume you have m observations, i.e., m data points in the dataset, where each data point contains:

- Values of features that take part in the prediction. These are the inputs. For example, previous 10 days' closing price and volume, represented by x_1 through x_{20} in Eq. 3.2.
- The observed value, represented by y in Eq. 3.2. This is the output.
- An implicit value of 1 for x_0.

The goal is to find the value of θ (θ_0 through θ_{20}) to minimize the error made by the model when applied to inputs from the observed historical dataset.

There are two popular methods for computing θ: *Normal Equation Method* and *Gradient Descent* Method.

3.4.1 Normal Method

You can think of the hypothesis being expressed as a matrix equation, shown in Eq. 3.4:

$$X * \theta = Y \tag{3.4}$$

where

X is matrix of dimensions (m, n + 1), m rows (one row per data point), and n + 1 columns (first column with implicit feature value of 1 and then n columns – one per feature)
θ is a column matrix of dimension (n + 1, 1), having values θ_0 through θ_n
Y is another column matrix of dimension (m, 1), having values Y_1 through Y_m, representing the observed values for the m data points

Before proceeding further, make sure that you understand Eq. 3.4 represented in matrix form.

While you want the solution to Eq. 3.4 to be as accurate as possible, in reality, it might not be possible to find the value of θ such that the equation would be satisfied exactly for each of the m observations. Hence, the attempt is to find θ such that $X * \theta$ is as close to Y as possible.

Modify Eqn. 3.4 into 3.5, where we replace Y by Y_p, which denotes the *predicted* value of Y:

$$X * \theta = Y_p \tag{3.5}$$

This now changes our problem definition slightly. You need to find θ such that Y_p is as close to Y as possible, i.e., $(Y - Y_p)$ should be as close to 0 as possible. And this should be evaluated on the basis of all the rows of data. Sometimes, a value of θ which improves the error for a given row could deteriorate the error for another row. So, you want to consider the total (or, average) errors across all the rows. When you consider the errors for all the rows, large positive error for a row could potentially be cancelled by large negative values on another row. To overcome this problem, you want to consider the square of an error, which is always positive, and hence, there is no chance of errors from different rows being cancelled by each other. Since Y, Y_p are both column matrixes, $Y - Y_p$ is also a column matrix, and squaring each number (to get rid of negative values) can be thought of as $(Y - Y_p)^T(Y - Y_p)$ where $(Y - Y_p)^T$ is transpose of matrix $(Y - Y_p)$. This is equivalent to squaring of each element of column matrix $(Y - Y_p)$.

So, now, the requirement is to minimize $(Y - Y_p)^T(Y - Y_p)$

Since $Y_p = X\theta$, the above requirement can be rewritten as minimize $(Y - X\theta)^T(Y - X\theta)$.

The term that you want to minimize is called *cost function* or the *loss function* and is often represented by $J(\theta)$. The θ in bracket indicates that the value of the cost function is dependent on the values of θ. Equation 3.6 gives the expression of our cost function that you want to minimize:

$$J(\theta) = (Y - X\theta)^T(Y - X\theta) \tag{3.6}$$

Differential calculus says that for any curve, at its minimum value, the slope of the curve will be 0. Using matrix differential calculus, the slope of the $J(\theta)$, i.e., *taking a derivative with respect to* θ, will be $2X^TX\theta - 2X^TY$. Equating this slope to 0, you will get Eq. 3.7:

$$\begin{aligned} 2X^TX\theta - 2X^TY &= 0 \\ X^TX\theta &= X^TY \\ \theta &= (X^TX)^{-1}X^TY \end{aligned} \tag{3.7}$$

So, given matrices X and Y for a large number of sample set, it is possible to determine the value of column matrix θ that will provide you with a way to predict a new value of y for a new set of features.

The only problem with Eq. 3.7 is that it needs matrix inversion. Sometimes, X^TX may not be invertible, or, since matrix inversion is a compute intensive operation, sometimes, it might be computationally intensive to invert X^TX. In such cases, we use another method called *Gradient Descent*.

3.4.2 Gradient Descent Method

You know that predicted value $y_p = h_0 = \theta_0 * x_0 + \theta_1 * x_1 + \ldots + \theta_n * x_n$, and the error for any given dataset is equal to $y_p - y$.

Since you have m data points, the total error over these m data points should be considered. And that total error is given by:

$$J(\theta) = \frac{1}{2m} \sum_{i=1}^{m} \left(h\left(x^i\right) - y^i\right)^2 \tag{3.8}$$

where

$h(x^i)$ represents the predicted value y_p for the ith dataset
$h(x^i) - y^i$ represents the error between the predicted value y_p and the actual observed value y for the ith dataset
m is the number of data points in the dataset (such as historical data used in modeling)
$J(\theta)$ is the total error term of the model for all the datasets

The squaring is done to make each individual error positive, in order to prevent mutual cancellation of positive and negative error values, when summed over a large number of dataset.

The summation symbol ($\sum$) is to add the errors for each dataset, so that you take any decision based on the total error, rather than the error for any single observation.

And the division by $2m$ is for getting the average. Strictly speaking, you should be dividing by m for getting the average. However, you will see later (in Sect. 3.4.2.4) that we will be using a scaling by an arbitrary constant anyways. And so, you can divide (or, multiply) by any constant. Division by 2 is for ease in some further mathematical formulation.

Since Eq. 3.8 for the cost function is quadratic, its curve would look something like shown in Fig. 3.1.

This shape of the curve gives you a few very interesting properties.

- There is only one minimum value on this curve.
- The further you are from the local minima, the higher is the absolute value of the slope.

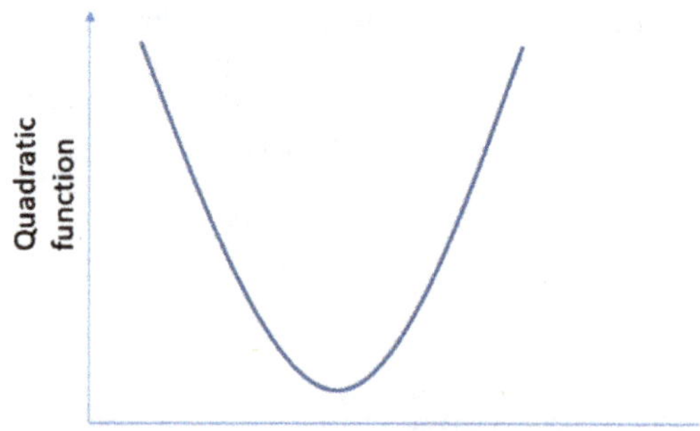

Fig. 3.1 General shape of a quadratic equation

- The positive and negative value of the slope will also indicate whether you are to the right of the minima or to its left.

The name *Gradient Descent* comes because the solution exploits these three properties of the quadratic curve in Fig. 3.1. The solution will *descend* on this curve toward the minimum, using gradient (or, slope) direction and the value as a guidance.

3.4.2.1 Determine the Slope at Any Given Point

The cost function $J(\theta)$ depends on coefficients θ_0 through θ_n, and you need to determine each of these θs. So, you need to determine the slope of J with respect to each θ_j, where j takes the value 0 through n. This concept of finding a slope with respect to only one variable (when the function is dependent on multiple variables) is called *partial derivative* and is denoted by ∂. So, $\frac{\partial J}{\partial \theta_0}$ represents slope of $J(\theta)$ with respect to θ_0.

You will notice that the expression for $J(\theta)$ has uniform representation in any θ_j for all values of j ($j = 0, 1, 2, \ldots, n$). Using differential calculus on Eq. 3.8, you will get partial derivative values represented by Eq. 3.9:

$$\frac{\partial J}{\partial \theta_j} = \frac{1}{m} \sum_{i=1}^{m} \left(h\left(x^i\right) - y^i\right) \cdot x_j^{\ i} \quad (\text{for } j = 0, 1, 2, \ldots, n) \tag{3.9}$$

where

Subscript $_j$ refers to the values corresponding to *j*th parameter
Superscript ^{i} refers to the values corresponding to *i*th dataset
x_j^i refers to the value corresponding to *j*th parameter in *i*th dataset

Thus, at any given point, it is possible to determine the slope of $J(\theta)$ with respect to each of θ_j ($j = 0, 1, 2, \ldots, n$).

Let us understand the expression forming the slope.

$h(x^i) - y^i$ is the error between the predicted value y_p and the actual observed value y for the *i*th dataset. Here, you retain the sign, since you want to know which side of the minima you are operating on. And this error is then multiplied by the value corresponding to the specific parameter of that dataset.

And the expression is summed over all the m datasets.

3.4.2.2 Initial Value

You can pick up any initial value for each of θ_j ($j = 0, 1, 2, \ldots, n$). It just does not matter what value you choose. Often, people start with a choice of 0 for each of θ_j ($j = 0, 1, 2, \ldots, n$).

3.4.2.3 Correction

Once you have a set of values of θ_j ($j = 0, 1, 2, \ldots, n$), use Eq. 3.9 to determine the slope of $J(\theta)$, with respect to each θ_j ($j = 0, 1, 2, \ldots, n$). And, now, update each of these θ_j. The way to apply the correction is:

$$\theta_j = \theta_j - \alpha . \frac{\delta J}{\delta \theta_j} \tag{3.10}$$

where

slope $\frac{\delta J}{\delta \theta_j}$ is obtained through Eq. 3.9
α is a constant and is a correction factor

So, effectively, if for any θ_j, the slope is positive, reduce the value of θ_j, and if the slope is negative, increase the value of θ_j.

For any given set of θ, once you have determined the slopes for each θ_j (using Eq. 3.9), and applied the correction (using Eq. 3.10), you now have a new set of θ, and this θ should hopefully be a better choice compared to the previous set.

In order to apply Eq. 3.10 for updating θ_j, you need to understand α which also appears in the equation, and we look at this variable in Sect. 3.4.2.4.

As you have seen, slope is used to determine the direction of correction as well as magnitude of correction. The idea is to reach the point of minimal error at the fastest speed. The rest of the book uses this concept to descend across the gradient (slope) of the error curve. However, there are some variations of algorithm for still faster descent. Chapter 18 shows some of those variations. Once you understand the usage of *Gradient Descent* for reaching your desired solutions, you can also decide to use one of the variations as explained in Chap. 18.

3.4.2.4 Learning Rate

Equation 3.9 gives you the slope. Its direction tells you whether to increase or decrease the θ_j. However, by how much? That "how much" is controlled by α, also called *learning rate*.

Consider the cost function curve shown in Fig. 3.2.

Assume that your current solution is at point A. In order to come to the minimal value, you want to go toward the left – an information that is given by the slope (positive value) determined at A. A high value of α will cause you to move significantly to the left, and you might now reach the point B, which is much closer to the desired solution. On the other hand, a smaller value of α will cause you to move only slightly to the left to reach point C. Hence, a smaller value of α will mean smaller steps, and hence it will take longer to reach the desired point.

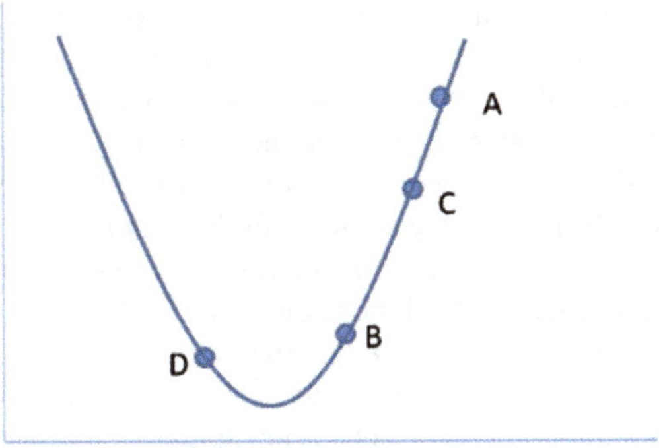

Fig. 3.2 Cost function curve and learning rate

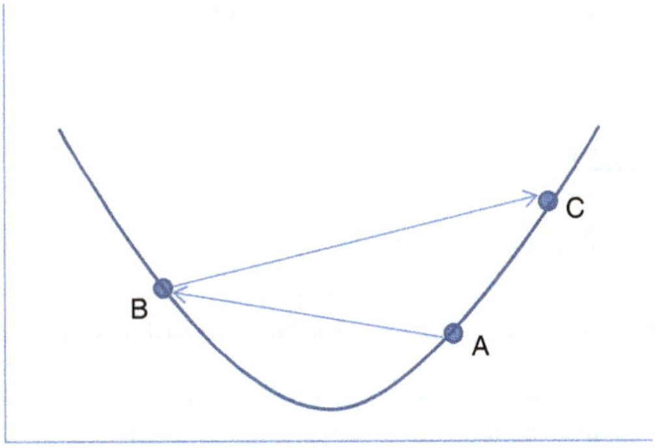

Fig. 3.3 High α causing divergence

On the other hand, a larger value of α would mean larger steps. While the obvious advantage is that it will reach toward the solution faster, there is always a risk that it might overshoot the solution. So, from *B*, at the next iteration, it could go to point *D*, which is on the other side of the desired solution.

The overshoot itself is not necessarily a problem, if at the next iteration, it tends to come back again toward the desired solution. However, with a high value of α, it is also possible that sometimes the solution does not *converge*; rather, with each iteration, the solution keeps on going further and further away from the desired solution. Figure 3.3 shows how the solution *diverges*, as it moves from *A* to *B* to *C*, due to a very high value of α.

So, a small value of α causes very small steps toward desired solution, while larger value of α causes large steps, and so it moves faster toward the solution but has the risk of overshooting or even diverging. Thus, you need to choose a good value of α.

Unfortunately, there is no good algorithmic way of finding the best value of α. You need to try various combinations and monitor if the solution continues to converge and how fast does it converge. So, try values of α in the range of 0.001 through 1, with various values in between, say 0.003, 0.01, 0.03, 0.1, 0.3, etc. Effectively, steps in multiple of ~3. If in doubt, choose a lower, rather than higher value.

Another interesting point is you would want to take larger steps, when you are further away from the solution, so that you don't have to take too many steps. And you would want to take smaller steps, when you are closer to the solution, so that you don't overstep. There is no need to change α for achieving this different step-size. This effect can be achieved due to slope changing. Given our cost function, the slope will be higher when you are further from the desired solution, and thus, you will be taking bigger steps. And closer to the desired solution, the slope would be smaller, and hence, you will be taking smaller steps.

This concept of α is providing a scaling factor for applying the correction to computed slope. Since there is anyways a final scaling applied to the slope, the initial arbitrary scaling of slope is not an issue, since an appropriate choice of α will nullify the arbitrarily scaled slope earlier. This is the reason why we had added an additional factor of 2 in the denominator in Eq. 3.8, which made the expression for slope easier.

3.4.2.5 Convergence

A solution is said to be *converging* when continued iterations of the computation bring the solution closer to desired value. Section 3.4.2.3 discussed how to apply correction to current value of θ, in order to get better values of θ. The questions are:

- How do you know you are done and don't need to continue applying corrections any further?
- How do you know the solution is even *converging*?

One good method is to keep monitoring the cost function $J(\theta)$. With each iteration, $J(\theta)$ should be decreasing. As long as $J(\theta)$ is decreasing, you know that you are descending on the cost curve, and hence, the solution is converging. On the other hand, if $J(\theta)$ is increasing, or even oscillating, that means either something is wrong in our algorithm/coding or the choice of α is too high and should be reduced. So, monitoring $J(\theta)$ can help you confirm that the solution is converging.

Now, you need to decide when to stop applying corrections and feel satisfied with the solution at hand. When the change in $J(\theta)$ is very small, you know that you are close to the solution. The cost function curve has shown that closer to the minima, the slope is very small, and hence, a very small change in $J(\theta)$ per iteration means you are close to the minimum. Depending on how close to the minimum do you want to go, you can choose the appropriate criterion for stopping to apply further corrections.

While determining the criterion for achieving closure, consider the value chosen for α. Assume you have chosen a very small value of α. That means the correction applied for each iteration would be very small, even if you are very far from the solution. So, a very small change in $J(\theta)$ may satisfy the criterion for your convergence. In reality, it is not that the solution has converged, just that your choice of α was so small that the iteration caused a very small step and hence, small improvement.

3.4.2.6 Alternate Method for Computing Slope

If you don't feel comfortable with *differential calculus*, this section provides an alternate method for computing the slope, at a given point, instead of Eq. 3.9 explained in Sect. 3.4.2.1.

You already have the current value of θ. You want to determine the slope of $J(\theta)$, with respect to each of the $\theta_0, \theta_1, \theta_2, \ldots, \theta_n$. Think of a very small number, ϵ. For determining the slope of $J(\theta)$ with respect to θ_0, determine the value of J, keeping all other parameters unchanged, except θ_0, which should be decreased by ϵ. Call this as J_1. Now, redo the same computation, except that θ_0 should be increased by ϵ. Call this as J_2.

Now, *slope* of $J(\theta)$ with respect to θ_0 is given by Eq. 3.11:

$$\frac{\delta J}{\delta \theta_0} = (J_2 - J_1)/(2 * \epsilon) \tag{3.11}$$

The expression on the right-hand side of Eq. 3.11 considers: when the parameter moved from $\theta_0 - \epsilon$ to $\theta_0 + \epsilon$, by how much did the cost function change and in which direction. This is what *slope* is about.

Apply the same mechanism for determining the slope of $J(\theta)$ with respect to each of the other parameters: $\theta_1, \theta_2, \ldots, \theta_n$. For determining the slope with respect to a single parameter, you need to determine the cost function twice, once with a slightly smaller value and once with a slightly larger value. So, you will need to determine the cost function $2n$ times for n parameters. And each computation of cost function is determined by summing of error square over all of the m datasets. It would be easier to do these computations as matrix multiplication, for quicker results.

This method for slope computation has been given only for the sake of conceptual clarity, if you are not comfortable with differential calculus. It would be good to get familiar with differential calculus way of computing slope, since some of the later chapters depend very heavily on slope computation, and computing slope through this method will unnecessarily be that much more compute intensive.

3.4.2.7 Putting Gradient Descent in Practice

This section provides the pseudo code to summarize what you have learnt so far for Gradient Descent. It starts with the assumption that you already have the dataset available.

Initialize all parameters to random values // θ initialized
Consider an α // Learning rate
While Error is reducing significantly {
  Compute slope of $J(\theta)$ with respect to each of $\theta_0, \theta_1, \theta_2, \ldots, \theta_n$ // using equation 3.9 (or 3.11)

Update each of $\theta_0, \theta_1, \theta_2, \ldots, \theta_n$ *// using equation 3.10*
Keep monitoring $J(\theta)$ *for convergence*

}

Note that you don't need to evaluate the actual cost function $J(\theta)$ for the Gradient Descent to work as we are computing change in cost function for adjusting θ. You may still want to compute it to observe/determine the convergence of the solution.

3.4.3 Normal Equation Method vs Gradient Descent Method

So, now you have two methods for applying Linear Regression. Which method should you use?

Obviously, Normal Equation Method is better, because it gives the exact solution in a single shot. On the other hand, Gradient Descent Method is an iterative process. Further, depending upon the choice of α, and the criterion for terminating the iterations, the solution arrived at would be different.

However, Normal Equation Method has one major disadvantage that needs matrix inversion. Not all matrices can be inverted, and even when invertible, very large matrices can take a long compute time to invert. A general rule of thumb could be, if you have the $X^T X$ (from Eq. 3.7) to be larger than 1000×1000 matrix, you may be better off with Gradient Descent method. Some of the reasons for matrices not being invertible could be:

- Some of the features are directly related, and thus, some columns in X are redundant. Chapter 12 on *Principal Component Analysis* will explain getting rid of redundant columns.
- There are more variables than data. This anyways means a non-unique solution. This should be fixed by having many more datasets, compared to the number of features that you have in your model.

In the case of Linear Regression, there is a closed-form solution, through Normal Equation (of matrices). However, for many problems that you will see in subsequent chapters of the book, such closed-form solutions may not be possible. Iterative methods based on Gradient Descent are used for solving those problems.

3.5 Logistic Regression

Logistic Regression provides a 0 or 1 prediction, rather than a real value for a continuous variable. To that extent, it is a *classification* technique and not a *regression* problem.

Consider a situation, where you want to predict if the next closing price of a stock (or index) will be higher than today's closing price. One way of doing this prediction is to use the Linear Regression technique (learnt in Sect. 3.4) to predict the next day's closing price. If this predicted price is higher than today's closing price, you can predict that the closing price will be higher. And if the predicted price is way higher, you can have a higher degree of confidence in your prediction of next closing price being higher than today's closing.

However, there is another better method of predicting the outcome; if the answer desired is of Boolean form, 0/1 or *yes/no*, then use Logistic Regression. Similarly, if you want to determine if the given picture is a human or not, you need a *yes/no* answer.

3.5.1 Sigmoid Function

Since interest is in predicting a *Boolean* answer, let us choose a function which operates within the range 0 to 1. *Sigmoid* is one such function that is extensively used for this purpose. Sigmoid function is defined as:

$$g(z) = \frac{1}{1 + e^{-z}} \tag{3.12}$$

Consider $z = \sum_{i=0}^{n} (\theta_i * x_i)$, and our hypothesis will be given by:

$$h_\theta(x) = g(z) \tag{3.13}$$

where

$g(z)$ is the Sigmoid function represented by Eq. 3.12
$z = \sum_{i=0}^{n} (\theta_i * x_i)$. By the way, $\sum_{i=0}^{n} (\theta_i * x_i)$ can also be represented as $\theta^T X$
x_0 is always considered 1, as in Linear Regression

Table 3.1 shows the values of $g(z)$ for various values of z (or, $\theta^T X$). It shows that the values of $g(z)$ lie within the range 0–1.

This $h_\theta(x)$ represents the probability that the predicted query (y) will be 1. If the predicted value of the hypothesis is above 0.5, you predict a value of 1, or else, predict a value of 0. As you can see, unlike Linear Regression predicting a continuous variable, Logistic Regression is predicting a binary class of 0 or 1.

So, now the main job is to find the appropriate values of θ_j, for $j = 0, 1, 2, \ldots, n$.

Table 3.1 Range of $g(z)$ values for various values of $\theta^T X$

z (or $\theta^T X$)	$g(z)$ (or, hypothesis value)
Very large negative value	~0
Below 0	Between 0 and 0.5
0	0.5
Above 0	Between 0.5 and 1
Very large positive value	~1

3.5.2 Cost Function

You could think of a cost function $J(\theta) = \frac{1}{2m}\sum_{i=1}^{m}\left(h(x^i) - y^i\right)^2$ as given by Eq. 3.8 to apply to Logistic Regression. However, now that our hypothesis (h_0) is no longer linear (Sigmoid is a nonlinear function), this cost function has multiple local minima and so is not very suitable. Thus, let us identify a cost function that is more suitable for the current purpose.

The cost function should be defined such that, for any given observed data, the value is very high when the hypothesis does not match the observation and the cost value is very low when the hypothesis matches the observation.

One such cost function (for a single data observation) is given by Eq. 3.14:

$$\text{Cost} = -y\log(h_0(x)) - (1-y)\log(1-h_0(x)) \tag{3.14a}$$

$$\text{Or,}\quad \text{Cost} = -[y\log(h_0(x)) + (1-y)\log(1-h_0(x))] \tag{3.14b}$$

Table 3.2 shows how this cost function fits our criterion of putting in higher cost to situations where the hypothesis does not match the actual observation and a low cost where the hypothesis matches the actual observation.

So, the overall cost function turns out to be:

$$J(\theta) = -\frac{1}{m}\left[\sum_{i=1}^{m} y^i \log\left(h(x^i)\right) + \left(1-y^i\right)\log\left(1-h(x^i)\right)\right] \tag{3.15}$$

where the superscript ^{i} denotes the ith dataset.

Notice that, this time, we are not using an additional factor of 2 while computing the average.

3.5.3 Gradient Descent

Now that you have the right cost function, you can apply the Gradient Descent algorithm. Slope of J with respect to any θ_j is given again by Eq. 3.9 (being reproduced):

Table 3.2 Cost corresponding to combinations of hypothesis and observed values

Y	$h(x)$	Part I $-y \log h_0(x)$	Part II $-(1-y)\log(1-h_0(x))$	Overall cost
$Y=0$	$h(x) = \sim 0$	0 (because y is 0)	~0 (because log (~1) is 0)	~0
	$h(x) = \sim 1$		High value (because negative of log ~0)	High value
$Y=1$	$h(x) = 0$	High (because negative of log ~0)	0 (because $1-y$ is 0)	High value
	$h(x) = 1$	~0 (because log ~1)		~0

$$\frac{\delta J}{\delta \theta_j} = \frac{1}{m} \sum_{i=1}^{m} \left(h(x^i) - y^i \right) \cdot x_j^{\,i} \text{ (for } j = 0,1,2,..,n) \tag{3.9}$$

While the equation for the slope is the same for Linear Regression as well as the Logistic Regression, note that the hypothesis $h(\theta)$ is not the same.

And similar to Linear Regression, once you determine the slope of J with respect to each θ_j, update the value of θ using Eq. 3.10 (being reproduced):

$$\theta_j = \theta_j - \alpha \frac{\delta J}{\delta \theta_j} \tag{3.10}$$

and keep iterating till you converge.

3.6 Next Steps

In this chapter, you learnt techniques to predict values for continuous variables and also to predict a *yes/no* kind of decision, along with confidence (probability) of the decision being correct. In the next chapter, you will learn techniques to further refine learnings from this chapter. Hence, it is good for you to be comfortable with concepts learnt so far.

- Try to predict the closing price of a stock or index based on last 10 days of closing prices and volumes.
  - Use both Normal Equation Method and Gradient Descent Method.
  - See if you get the same answer using both methods.
  - See how your answers evolve as you play with different values of α and criterion for determining convergence.
- Try to predict whether the next closing price will be higher than 2% of the previous closing price (Logistic Regression).

Use any tool/language of your choice. Wherever possible, try to see the problem in matrix form and apply matrix-based computations.

Hint for arranging data:

Choose a stock/index of your choice, and collect (for many days – in the order of 100s) closing values and volume for that stock. For any day, the closing price of that day is y for that dataset, and the last 10 days' closing prices and the volumes are the x_1 through x_{20} features for that dataset. x_0 will be 1 for each dataset. Since you have collected the closing prices and volumes for many days, you should be able to construct the x and y dataset for those many days. Now, apply the algorithms.

By formulating the problem of stock price prediction based on prior 10 days' data, we are not claiming that this will give the right prediction. If you want to trade based on ML, don't depend on just 10 days of data. Reason for suggesting a problem based on stock/index is because this data is very easily available. So, you can concentrate on learning the algorithm, rather than on hunting for the data.

3.7 Key Takeaways

Besides learning how to create a prediction model, both for obtaining values and for *yes/no* type decision, another major learning from this chapter is that even if you don't have a closed-form solution, you can arrive at a solution by systematically iterating. This is a concept that will be used in many subsequent chapters.

Chapter 4
Improving Further

In this chapter, we will further refine the techniques learnt in the previous chapter. With these refinements, you will be able to achieve one or more of the following:

- Reach convergence faster
- Get more accurate prediction
- Consider higher order (or other nonlinear) contributions of the participating features

Most of the concepts explained in this chapter apply both for *Linear Regression* and *Logistic Regression.*

4.1 Nonlinear Contribution

Suppose, you believe that a specific feature's (say, x_3) contribution is actually the square of its value or that the product of two features (say, x_3 and x_4) is more meaningful in terms of predicting the value of interest. You may want to consider any such variation of the features or combination of features. So, now your hypothesis function gets modified to Eq. 4.1:

$$y = \theta_0 * x_0 + \theta_1 * x_1 + \theta_2 * x_2 + \theta_3 * {x_3}^2 + \theta_4 * x_3 * x_4 + \ldots. \tag{4.1}$$

If you now consider ${x_3}^2$ and $x_3 * x_4$ as features by themselves, you will get to the original equation form, as shown by Eq. 3.2, except that you would have additional features which are the values of ${x_3}^2$, $x_3 * x_4$, and whatever other combinations you desire to consider.

Even if you decide that the prediction is a function of square (or whatever power you desire) of a combination of all the features, even that can be expressed as a linear hypothesis as shown in Eq. 4.2:

G. Rebala et al., *An Introduction to Machine Learning*,
https://doi.org/10.1007/978-3-030-15729-6_4

$$y = \theta_0 * x_0 + \theta_1 * x_1 + \theta_2 * x_2 + \theta_3 * {x_1}^2 + \theta_4 * {x_2}^2 + \theta_5 * x_1 * x_2 + \ldots. \quad (4.2)$$

Notice that you are including different powers of individual variables (or their combinations); however, the overall hypothesis is still a linear function of θs. You can also extend this concept to consider other function of any feature, such as *trigonometric* or *logarithmic* values.

While the features themselves have been considered with higher order, understand that the cost function is still quadratic and, hence, the rest of the discussion of the previous chapter still applies, as long as you consider the derived features as if they were independent features by themselves.

Similarly, for Logistic Regressions, you can derive values from primary features and use these derived features as a feature by itself.

This simple creativity of adding derived features allows you to exploit Linear Regression and Logistic Regression to be driven by nonlinear contribution of individual variables.

Clearly, as you consider higher order powers, the number of parameters increases drastically, and that has an effect on your computation efforts.

4.2 Feature Scaling

The variables considered in modeling for regression represent *features*. There can be significant number of features in creating the model. Some of the features may have a significant difference in their range of values. For example, individual stock closing price could be in two or three digits, while volume might be in millions. And then, x_0 is always kept 1. The algorithm reaches convergence much faster, if the feature values are of the same order. One good method to scale the values to be of the same order is to modify each feature value (except x_0) using Eq. 4.3:

$$\text{Updated } \left({x_j}^i\right) = \frac{\text{Original } \left({x_j}^i\right) - \mu(j)}{\text{Range } (j)} \quad (\text{for } j = 1, 2, \ldots, n) \quad (4.3)$$

where

x_j^i refers to the value corresponding to *j*th parameter in *i*th dataset

$\mu(j)$ refers to mean of all values of *j*th parameter considering all *m* dataset (i.e., *j*th column of *X* matrix)

Range(j) refers to the (*max* - *min*) for all values of *j*th parameter (considering all *m* dataset)

Notice that this feature scaling is not applied to x_0. An attempt to modify x_0 using Eq. 4.3 will cause the column to become all 0s.

The modified data table (for features), after applying feature scaling, through Eq. 4.3 will have each value (including x_0) within the range -1 to $+1$ (both sides inclusive).

It is not necessary that you have to put the values strictly in the range -1 to $+1$. The main requirement is to have the values within similar scale. So, instead of *range*, you can also use *standard deviation* in Eq. 4.3 for updating the values. This might cause the values to go in the range of ~ -3 to $\sim +3$, and that is OK. Similarly, instead of $\mu(j)$, you can use an indicative value around which most of the values of *j*th parameter lie. This may cause the values to be skewed on one side (e.g., -3 to $+1.5$), and that is also OK.

Avoid scaling the values by a very large denominator. Such scaling would cause all the values to lie in a very small range (e.g., -0.000005 to $+0.000005$). If you do this kind of scaling, the values will differ from each other by such a small amount that the difference will be lost within the precision of your computer system.

Theoretically, feature scaling is not needed for *Normal Equation*-based solution (Eq. 3.7); however, if the feature values are large compared to x_0 (which is always 1), the matrix inversion often has a large error introduced due to numerical precision issues. A good way to check if your dataset is suffering from this problem even for Normal Equation, is to multiply $(X^TX)^{-1}$ by (X^TX). If you get a matrix which is very close to *Identity Matrix*, you can safely use *Normal Equation*, without having to worry about feature scaling, else consider feature scaling for Normal Equation method also.

4.3 Gradient Descent Algorithm Variations

The Gradient Descent algorithm explained in Chap. 3 is a *Batch Processing* algorithm. The algorithm considers all the data together in each processing step. Notice Eq. 3.8 (Sect. 3.4.2) for finding the cost (i.e., error) and Eq. 3.9 (Sect. 3.4.2.1) for finding the slope. Both these equations need data to be computed for each of the *m* data points. Since machine learning training involves a large amount of data, that means a lot of data points will need to be accessed each time. And that means access to secondary storage many times – for every iteration of θ computation. Minor variations to this *Batch Processing* algorithm to avoid accessing the whole dataset so often will optimize processing time for the solution. Before you go to the variations in algorithms, it would be good to understand the concept of *Cost Contour*.

4.3.1 Cost Contour

Figure 4.1 shows a contour diagram for cost. For ease in understanding, consider only two variables θ_0 and θ_1.

This diagram shows how total cost, $J(\theta)$, varies with different combination of values for θ_0 and θ_1. Each closed loop represents a specific value of $J(\theta)$. So, points *A* and *B* represent the same cost; similarly, points *C* and *D* represent the same cost.

Fig. 4.1 Contour diagram

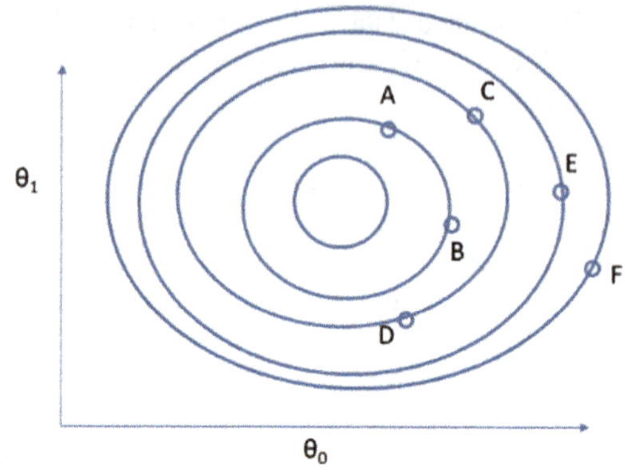

The points on inner loop represent lesser cost. So, *A* represents lower cost compared to *C*, which represents a cost lower than *E*, and so on. The whole point of Gradient Descent algorithm is to go from current cost loop toward inner loop.

Batch Processing algorithm allows to scan the whole contour (since the entire dataset is being processed) and chose a path that would be steepest at that point. For example, if the current values of θ_0 and θ_1 are such that $J(\theta)$ is at point *E*, the next iteration will tend to take the shortest path toward the innermost circle. It may not reach the innermost circle in a single shot, but its direction will be toward the innermost circle.

4.3.2 Stochastic Gradient Descent

In Sect. 3.4.2.1, you saw the computation of the slope given by Eq. 3.9 (being reproduced):

$$\frac{\partial J}{\partial \theta j} = \frac{1}{m} \sum\nolimits_{i=1}^{m} \left(h(x^i) - y^i\right) \cdot x_j^{\ i} \text{ (for } j = 0,1,2, \ldots, n) \tag{3.9}$$

In this variation of the algorithm, you consider the slope based on only a single data point, selected *stochastically* (i.e at random) from the dataset, as given by Eq. 4.4:

$$\frac{\partial J}{\partial \theta j} = \left(h(x^i) - y^i\right) \cdot x_j^{\ i} \text{ (for } j = 0, 1, 2, \ldots, n) \tag{4.4}$$

where

Superscript ^{i} refers to the values corresponding to *i*th data point, which is the current single data point under consideration

$x_j^{\ i}$ refers to the value corresponding to *j*th feature in *i*th (i.e., the current) data point

And then, based on Eq. 3.10 (being reproduced), you update the values of θ, based on this single data:

$$\theta_j = \theta_j - \alpha \cdot \frac{\partial J}{\partial \theta_j} \quad (\text{for } j = 0,1,2, \ldots, n) \tag{3.10}$$

where slope $\frac{\partial J}{\partial \theta j}$ is obtained through Eq. 4.4. Notice that here the slope is being determined by only the current data, rather than looking at all the data. You will continue to further adjust θ, based on the value of slope determined by the next data and so on. Hence, each computation of slope is very fast.

Clearly, the last data has the final say in which way should the slope be moving. So, repeat the whole algorithm multiple times, i.e., after a single pass through all the data, make a few more passes through the same set of data again. This reduces the higher degree of influence that the final data may exercise. One more exercise that you should consider is to randomly shuffle the entire dataset, so that the order is not dependent on the order in which you had tabulated the original data.

This entire algorithm can be explained by the following pseudo code:

```
Initialize θ  // Random initial value
Consider α // Learning Rate
Shuffle entire data
For K = 1 to N {
    For I = 1 to m {
        Compute slope based on I^th data // using equation 4.4
        Update θ //using equation 3.10
    }
}
```

The inner loop goes over each data and keeps updating θ based on each data. Thus, each data has an opportunity to influence the value of θ. And, the outer loop iterates over the same set of data multiple times. The outer loop serves two major purposes:

- If the amount of data (and hence the number of times θ got updated) is not enough to reach the correct value of θ, the outer loop provides additional opportunities for θ to be updated.
- It dampens the influence of the last data. For example, if the last data value wanted a θ which was totally different from what other values wanted, this last value would shift in the incorrect θ direction. However, due to next iteration of the outer loop, the other values will once again get a chance to bring θ back to where it should be.

This also gives an indication of the number of times the outermost loop should be looped through. If the number of data points (i.e., m) is large enough, the innermost loop itself will provide enough opportunities for updating θ that it will reach the desired value. In that case, a single iteration of the outer loop is sufficient.

If not, you should iterate multiple times for the outer loop. Maybe, iterate up to about a dozen times.

Looking from the perspective of Cost Contour diagram, Batch Processing algorithm took steps toward steepest descent from its current position. And hence, it needed lesser number of steps to reach toward minima.

On the other hand, Stochastic Gradient Descent algorithm looked at only one data at a time and took a path based on that one data. Hence, it moved toward the lower cost, but not necessarily in the steepest path. Hence, it might take more number of meandering steps to reach toward the local minima. It still saves total compute time, because the computation for each step is very small (and hence fast). The time saved is not just in terms of compute but also in terms of not having to access a large amount of data from secondary memory devices. Chapter 18 provides some suggestions on reducing the meandering – by considering prior slopes also. So, you can enjoy faster iteration as well as reduced meandering (thus, shorter path towards the final solution).

Sometimes, it is possible that even after multiple iterations of the outer loop, the solution may not reach the global minima. The values of θ may just keep going round in a contour that is very close to minima, but not the real minima. In most cases, that should still be close to the global minima, close enough to be considered as converged to a solution.

4.3.2.1 Convergence for Stochastic Gradient Descent

As the algorithm is processing each data point, you might want to monitor that the algorithm is converging toward a solution. For monitoring convergence, define your cost function as given in Eq. 4.5:

$$J(\theta) = \frac{1}{2} \left(h\left(x^i\right) - y^i\right)^2 \tag{4.5}$$

where superscript ^{i} refers to values corresponding to ith data point, which is the current data point under consideration.

Determine the cost before applying the correction to θ based on the slope for this specific data. As you plot $J(\theta)$ at each iteration, this cost curve should generally be going down in value. Notice that the cost is being determined based on a single data value, rather than the whole dataset. Plus, at each iteration, a different data point is being used to compute the cost.

Thus, unlike Batch Gradient Descent method, where the cost curve shows a decrease for each iteration, here, the cost curve might show an occasional increase. This occasional increase represents the situation of a meandering path toward the solution, rather than the shortest path.

If you desire, you may still compute the cost value as given in Eq. 4.5, average the cost over 100 iterations, and plot this value; and then for the next 100 iteration, plot the average of these 100 iterations, and so on. That is, you compute the average cost for every 100 data considered, and plot this average cost. Here also, you should see a generally falling curve with an occasional increase. Due to larger dataset being considered, this curve is expected to be a bit smoother, and hence, you should see a fewer number of increases in the plot for the cost curve. If, instead of considering the average for 100 data, you consider a larger number (say, 1000), the plot is expected to be still smoother.

The simplest is to just plot the cost for each iteration and just be aware and accept that the curve will show a lot of instantaneous ups and downs, with a generally downward direction.

4.3.3 Mini Batch Gradient Descent

Stochastic Mini Batch Gradient Descent is a hybrid approach between Batch Gradient Descent and Stochastic Gradient Descent.

In Stochastic Mini Batch Gradient Descent, the whole dataset is divided into *N* groups (also called mini batches), where data points in each group are chosen at random. Each group would have *m/N* data points.

Now, you apply the Gradient Descent algorithm, but, for any iteration, determine the slope solely based on the current group under consideration. So, the equation for the slope would be as given in Eq. 4.6:

$$\frac{\partial J}{\partial \theta j} = \frac{1}{m/N} \sum\nolimits_{i=1}^{m/N} \left(h(x^i) - y^i\right) \cdot x_j^{\,i} \quad (\text{for } j = 0, 1, 2, \ldots, n) \tag{4.6}$$

where

superscript ^{i} refers to the index of data points in the current group under consideration

N is the number of groups of data points

m is the total number of data points

Now, update the value of *J(θ)*, based on the slope determined as per Eq. 3.10:

Initialize all parameters to whatever values // θ initialized
Consider an α // Learning rate
Shuffle entire data
Divide the whole data set into N groups // each group containing m/N data
Iterate while solution is not converged {
For K = 1 to N {
Compute slope of J(θ) with respect to each of θ_0, θ_1, θ_2, …, θ_n // using equation 4.6 for data within group K
Update each of θ_0, θ_1, θ_2, …, θ_n // using equation 3.10
}
} // solution converged

So, for each iteration, the slope is computed based on only *m/N* data points (rather than *m* data points for Batch Gradient Descent or a single data point for Stochastic Gradient Descent). Besides the compute time savings (compared to Gradient Descent), there are also major savings in not having to access the secondary memory for each iteration. If your hardware resources support parallel processing for vectorized implementation, you can perform the computation for each iteration in parallel, making each iteration take almost the same time as Stochastic Gradient. In such cases, this algorithm will perform faster than Stochastic Gradient, since the number of iterations would be lesser for this implementation.

For determining *N* (the number of groups), choose *N* such that each group has as many data points, as can be supported in parallel by your hardware resources. If your hardware resources (or software tools) do not support parallel implementation of Batch Processing, choose *N* so that each group contains a constant number of points. Pick the constant that works best for your hardware (e.g., 10).

4.3.4 Map Reduce and Parallelism

Suppose you have access to parallel computing environment (such as FPGAs, GPUs, etc.), you can make additional improvement to your algorithm. Say, you have access to *N* compute resources, which can be utilized in parallel. Once again, divide your entire dataset into *N* groups. Send one group of data points to each machine, and compute the slope individually for each of these *N* groups. Consider the average of these slopes as the final slope, and now update θ based on this average slope.

There is a very subtle difference between Mini Batch Gradient Descent and Map Reduce-based Gradient Descent. In both algorithms, the entire dataset is divided into *N* groups, and slope is determined based on dataset of one group.

In Mini Batch, the value of θ is updated based on the slope determined for each group. So when there are *N* groups, there are *N* updates to θ. In Map Reduce, the individual slopes calculated for each group are averaged across all the *N* groups, and then θ is updated based on this average slope. So there is a single update (per iteration) for θ. The slope computation equation for any group is still the same (from Eq. 4.6) in both the methods.

The following pseudo code shows the Map Reduce algorithm:

```
Initialize all parameters to whatever values // θ initialized
Consider an α // Learning rate
Divide the whole data set into N groups // each group containing m/N data
Iterate while solution is not converged {
  For K = 1 to N {
    Compute temp slope of J(θ) with respect to each of θ0, θ1, θ2, ..., θn // using equation 4.6 for data within group K
  }
  Average the temp slopes obtained over N groups
  Update each of θ0, θ1, θ2, ..., θn // using equation 3.10
} // solution converged
```

4.3.5 *Basic Theme of Algorithm Variations*

As you looked at the various algorithms, a few things are common.

- The updates to θ based on slope and learning rate
- Iterations till solution converges

The only fundamental difference is the way the slope is being computed. The Batch Gradient Descent algorithm looks at the entire dataset and takes a global view of Cost, and thus, each step is in the direction toward global minima. However, it takes too long to compute each step.

Other algorithms look at only a subset of data at a time and try to go toward minimum based on that subset of data. Thus, each step is in the right direction but based only on the data being considered. Considering the entire dataset, the direction could potentially be different. The aim of these algorithm variations is to determine each step quickly, even if it is not strictly in the direction of the global minima. And, if available, make use of parallelism to determine the step quickly.

Since each step is looking only at a subset of the data points, it is possible that some of these variations may not reach global minima, and you may just move around close to the minima. You can create your own variations of the algorithm, as long as you have understood and appreciate all these variations.

One nice modification could be to use one of these methods to reach close to the minima rather fast. Then use the Batch method of Gradient Descent for the last few steps toward the global minima.

The method variations described in this section have used Linear Regression and Logistic Regression for explaining. However, the same concepts apply for other forms of machine learning problems that we will be discussing in future chapters. It is important to know that there is no guarantee of reaching global minima for more complex Nonlinear Regression problems discussed in later chapters.

4.4 Regularization

You are using Regression (both Linear and Logistic) to find values for $n + 1$ parameters (including θ_0). You are aware that solving for $n + 1$ variables needs $n + 1$ equations. However, you are using m data points ($m \gg n$) to find the best values for these $n + 1$ variables. Thus, the result is not an exact solution, rather a solution which provides the best fit. Assume that you have considered features which do not really play any role in determining the outcome. However, since the algorithm is trying to do a best fit, it will determine the contribution of this feature also. On the face of it, there is nothing wrong with the process, if that contribution is able to better correlate with the dataset. The only trouble is, while this contribution can provide a good correlation with existing data points, it might be unnecessarily playing a role in predicting the new values, when in reality, it should not be influencing the prediction.

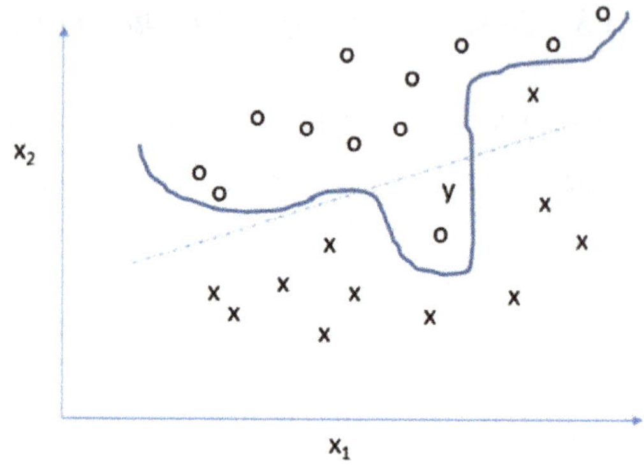

Fig. 4.2 Complex curve to best fit the data

Look at Fig. 4.2, which explains artificially fitting the best curve.

Say, markers *x* and *o* represent 1 and 0, respectively – the two different decision classes of a classification problem – and you want to use Logistic Regression to identify the decision boundary between them. This decision boundary will be used to make the decisions for new data points. The dashed line represents a reasonably good classification boundary. However, it causes an incorrect decision for 2 points. One observation of *x* lies above the dashed line, and one observation of *o* lies below the dashed line. Given the right number of variables/features, the algorithm could very well come up with a decision boundary as shown by the curved line. As you can see, this decision boundary satisfies all the observed data, without any exception.

This curved decision boundary seems better than the simple (dashed) line in that it fits all the observations. However, it suffers from the following two major defects:

- This boundary is extra complex. While the line is purely linear, the curved decision boundary is making use of higher orders of the feature variables.
- This boundary makes so much effort to fit the existing data points that even a few incorrect data points could cause this boundary to be modified significantly and incorrectly.

In the same figure, consider an additional point represented by *y*. Say, after the decision boundary is identified, you need to predict the class for this new point. Based on the dashed line being the decision boundary, the prediction will say 1 (i.e., similar class as points represented by *x*), while the curved line being the decision boundary, the prediction will say 0 (i.e., similar class as points represented by *o*). Based on visual inspection, the choice of 1 seems to be a better decision.

What you are seeing here is an erroneous result in prediction, because of high effort made to fit in the prior data. This problem of fitting to data including anomalous data points is called *overfitting*. This causes us to introduce a concept called *regularization*. This concept allows you to simplify the models, if needed, for better prediction. This will allow you to get rid of or reduce the impact of unnecessary complexity in your models.

Define a regularization parameter, λ. This parameter adds a cost for the values of θ chosen, for each of $\theta_1, \theta_2, \ldots, \theta_n$. Note that the parameter does not add any cost for θ_0.

So, your cost function gets modified to Eq. 4.7.

$$J(\theta) = \frac{1}{2m}\left(\sum_{i=1}^{m}\left(h(x^i) - y^i\right)^2 + \lambda\sum_{j=1}^{n}\theta_j^2\right) \tag{4.7}$$

where

the first term is the same as the regular cost function as seen in Eq. 3.8
the second term is the additional term added, to add cost for each value of θ_j, except θ_0

When λ is large, the second term could have a high contribution, and so, θ_j (*for* $j = 1, 2, \ldots, n$) values are kept small, and so the overall model will be simpler. In the extreme case, for very high values of λ, only θ_0 survives, thus creating an extremely simple model, which predicts the same value always, irrespective of the values of other features. On the other hand, a small value of λ will cause the second term's contribution to be low, and hence, θ_j will be determined mostly by the first term. Eq. 3.8 can be seen as a special case of Eq. 4.7, where λ is 0.

So, effectively, λ can be used to tune the trade-off between the model simplicity and trying to fit all data. λ is not a parameter to be solved for but is actually a constant decided during tuning. So it is called a *hyperparameter*.

With Eq. 4.7 representing the cost function, the slope equation is given by:

$$\frac{\partial J}{\partial \theta j} = \frac{1}{m}\sum_{i=1}^{m}\left(h(x^i) - y^i\right)\cdot x_j^i + \frac{\lambda}{m}\theta_j \quad (\text{for } j = 1, 2, \ldots, n) \tag{4.8}$$

And, in each iteration, θ_j is updated using Eq. 3.10 (being reproduced):

$$\theta_j = \theta_j - \alpha \cdot \frac{\partial J}{\partial \theta_j} \tag{3.10}$$

Replacing the slope by the expression given in Eq. 4.8, and some reordering, you will get:

$$\theta_j = \theta_j\left(1 - \alpha\frac{\lambda}{m}\right) - \alpha\frac{1}{m}\sum_{i=1}^{m}\left(h(x^i) - y^i\right)\cdot x_j^i \quad (\text{for } j = 1, 2, \ldots, n) \tag{4.9}$$

Effectively, the second term in the equation for updating θ_j remains the same as what you had seen in Chap. 3. Only the first term has been modified. The numerical value of θ_j is diminished slightly, and then the correction is applied as per the slope and the learning rate, the same as the regular Gradient Descent algorithm. For θ_0 directly apply the correction, without diminishing its value.

4.4.1 Regularization for Normal Equation

If you are determining value of θ using Normal Equation, the Eq. 3.7 needs to be modified as:

$$\theta = \left(X^{\mathrm{T}}X + +\lambda I'\right)^{-1} X^{\mathrm{T}} Y \tag{4.10}$$

where

I' is a variation of Identity Matrix I. It is an $(n + 1)$ by $(n + 1)$ matrix. All non-diagonal elements are 0. The first diagonal element is 0, and all other diagonal elements are 1.

This ensures that the first element of column θ, i.e., θ_0, remains unaffected, while other elements of the θ are subject to slight reduction.

4.4.2 Regularization for Logistic Regression

In Chap. 3, you had seen that the equation for updating θ for Logistic Regression was the same as that for Linear Regression. Only the expression for hypothesis was different. The same concept holds with regularization also. Equation 4.9 can be used for updating θ_j (for $j = 1, 2, \ldots, n$). And for θ_0, there is no need to diminish it before applying the correction.

For the sake of completeness, with regularization, the cost function equation given by 3.15 is modified to:

$$J(\theta) = -\frac{1}{m}\left[\sum\nolimits_{i=1}^{m} y^i \ \log\left(h\left(x^i\right)\right) + \left(1 - y^i\right) \ \log\left(1 - h\left(x^i\right)\right)\right] + \frac{\lambda}{2m}\sum\nolimits_{j=1}^{n} \theta_j^2 \tag{4.11}$$

4.4.3 Determining Appropriate λ

You now have a way to find a good trade-off between model simplicity and accuracy for fitting current data. This trade-off depends on the appropriate choice of λ.

A good way to determine the most appropriate value of λ is to use different values of λ, and determine θ, for each choice of λ. And then, see which of these values of θ is most accurate in predicting new data. It is very important to test the effectiveness of θ on new data and not on the data which was used to determine θ in the first place. You can try various values of λ in the range 0–10, in steps such as 0, 0.01, 0.02, 0.04, 0.08, 0.16, etc.

4.4.3.1 Cross Validation

Since you started with m data randomly, choose around 70% of the data in one set, called training set, and the other 30% in another set called cross validation set. Let the respective counts be denoted by m_{test} and m_{cv}. The choice of which specific data goes into the training set and which one goes to cross validation set should be random.

Try different values of λ. For each value of λ, find the value of θ based on the training set data. For each of these θ, compute the error for the cross validation set. And, the value of θ that gives minimal error on the cross validation set is the most appropriate θ, and the corresponding λ is the most appropriate regularization parameter. The cross validation data effectively is a fresh set of data against which the hypothesis is being validated. This is why it should be distinct from the training data.

For the purpose of comparing the effectiveness of the hypothesis, you need to consider error on the cross validation set. And, this error should be computed based on the original cost function as given by Eq. 3.8 for Linear Regression or through the average number of misclassifications for Logistic Regression.

This concept of cross validation can be used for tuning of other hyperparameters also, for various ML algorithms that you will see in the subsequent chapters.

4.4.3.2 *K*-Fold Cross Validation

You can also use *K*-fold Cross Validation mechanism to identify the most appropriate choice of λ. Divide your data into K portions (or folds). Use the data in the first fold as cross validation data and the data in the remaining $K - 1$ folds as training data. Next, use the data in the second fold as the cross validation set and the remaining data as the training set and so on, till the data in each fold has been considered as cross validation once. Figure 4.3 explains the concept, using fivefold as an example.

4.4.4 Comparing Hypothesis

The concept of testing a hypothesis on a fresh set of cross validation data is very useful, not just to determine the value of λ but also to compare hypothesis. Suppose, you are not sure if a specific feature should be considered just by itself or its square should also be considered. Create two hypotheses, one with the feature's square and one without the square. Obtain the θ, corresponding to each hypothesis; and test the hypothesis on the cross validation set. The hypothesis which provides a better result on cross validation set is the better choice.

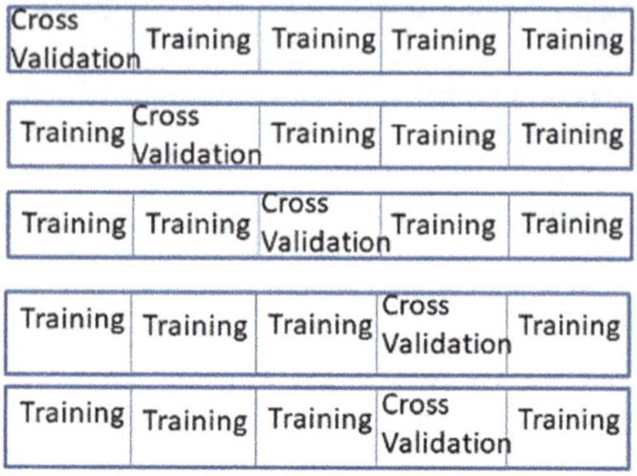

Fig. 4.3 *K*-fold cross validation

4.5 Multi-class Classifications

Logistic Regression explained in Chap. 3 on how to classify something into one of the two classes. Sometimes, you may need to classify something among more than two classes. Object detection is a very common example scenario of multi-class classification. Consider a situation where you may need to classify something among three classes.

4.5.1 One-vs-All Classification

The requirement to classify an object among one of the three classes can be thought of as three different classification problems:

- Group I vs everything else (i.e., Groups II and III)
- Group II vs everything else (i.e., Groups I and III)
- Group III vs everything else (i.e., Groups I and II)

Determine the *decision boundary* for each of these classification problems independently. Using three classification models, you will have three different predictions. These three prediction algorithms will have the same set of features but will have different values for the parameters (θ). Recollect that for Logistic Regression, the sigmoid function does not just give a 1/0 answer, but it provides the probability of the answer being 1.

Now, when new data comes, subject it to all three of the classification problems, and each of these will give the probability that this data corresponds to Groups I, II, and III, respectively.

The category or group that has the highest probability wins.

For example, assume that you were to train your machine learning algorithm to predict the possibility of:

- >5% gain
- >5% drop
- Range bound (i.e., movement within 5% range)

So, you will determine θ corresponding to >5% gain or less than 5% gain, and another θ corresponding to >5% loss or less than 5% loss and yet another θ corresponding to movement within 5% or more than 5%. For any new data, you will be able to get the probability for each of the three situations.

4.5.2 *SoftMax*

For multi-class classification, one of the popular approaches is to use *SoftMax* algorithm. Recollect that for Logistic Regression, you need to first determine $z = \sum_{i=0}^{n} (\theta_i * x_i)$, and, then, subject this z to a *sigmoid* operation.

In SoftMax classification, for each class, we determine the probability of the object belonging to that class. While one-vs-all will provide a final result of the object belonging to a class, SoftMax provides the probability of the object belonging to each class. Hence, instead of an all-or-nothing belongingness to a class, this algorithm provides a softer gradual answer.

4.5.2.1 Basic Approach for SoftMax

Given the input data point, determine the parameter z for each of the C classes. Determine $t = e^z$ for each of these C classes. Sum up the values of t thus obtained over all C classes. Divide t for each class by this sum. The resultant number denotes the probability of the input data point labelled as the corresponding class.

Table 4.1 explains this approach for three classes. Assume that the values of z shown in the table are already determined through $z = \sum_{i=0}^{n} (\theta_i * x_i)$. The last column of the table gives the probability of the respective class. The sum of this column should be 1. It is slightly higher in this example, because we have computed only to two decimal places.

Table 4.1 SoftMax computation explained

	z (determined)	$t = e^z$	t/Sum(t)
Class I	2	7.39	0.73
Class II	1	2.72	0.27
Class III	−3	0.05	0.005
Sum		10.16	1.005

4.5.2.2 Loss Function

The loss function to be minimized during the training phase is given by Eq. 4.12:

$$\text{Loss} = \sum_{j=1}^{C} -y_j \log\left(y_{\text{predicted}}\right) \tag{4.12}$$

where

C represents number of classes being considered
y_j represents the actual value of class j and, hence, can only be 1 or 0

Note that for an input data point, only the class it belongs to contributes to the loss function, since the multiplier y is zero for all other classes.

4.6 Key Takeaways and Next Steps

In this chapter, we further refined the algorithms learnt in Chap. 3. These refinements allow us to consider nonlinear contribution and consider trade-offs between time per iteration and number of iterations – to reach final solution faster, evaluate among multiple alternative hypothesis/models, and provide multi-class classification.

- Use your problem of Linear Regression from Chap. 3, include quadratic power for a few variables, and evaluate if that improves the forecast accuracy.
- Try at least two methods for Gradient Descent to get a feel of how convergence happens.
- Try to predict whether the next closing price will be a major gain or major loss or range bound (multi-class classification).

Chapter 5
Classification

Classification refers to the problem of identifying the category to which an input belongs to among a possible set of categories. The possible set of categories are labelled, and models are generally learned from training data. Classification models can be created using simple thresholds, regression techniques, or other machine learning techniques like Neural Networks, Random Forests, or Markov models.

Classification is a supervised learning algorithm where a training set of correctly identified or labelled data is available. The model learned from training data to identify the category or class of the input feature or data is called *classifier*.

The *classifier* can be a binary classifier or a multi-class classifier. A binary classifier identifies the input as belonging to one of the two output categories. For example, the mail received is a spam or not spam. A multi-class classifier identifies the input vector as one of more than two categories. For example, the mail received is a promotional email that represents some kind of advertisement, personal email received from friends or associates, or a spam email.

5.1 Decision Boundary

A *classifier* in a binary classification problem can be considered as representing a *decision boundary* or decision surface that partitions underlying vector space (i.e., the set of input feature vectors) into two classes. The *classifier* will classify all the points (corresponding to input feature vectors) on one side of the boundary as one class and other side of the boundary as another class.

A multi-class classifier partitions the input vector space into multiple sets with each set representing a class. Section 4.5 explains multi-class classification techniques. There is one more method for multi-class classification. This involves a combination of multiple binary classifiers that provide the decision boundary for each binary classifier independently. And, you can use intersections of these set of decision boundaries to identify the class of the input feature. For example, to classify

G. Rebala et al., *An Introduction to Machine Learning*,
https://doi.org/10.1007/978-3-030-15729-6_5

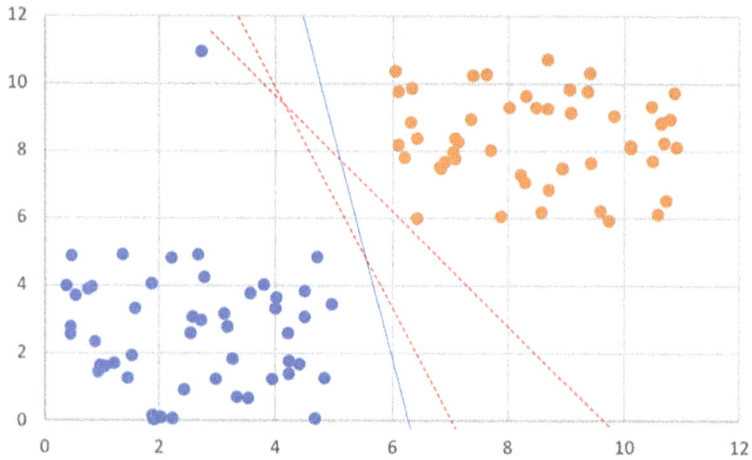

Fig. 5.1 Decision boundary for possible binary classifiers

email as personal, official, promotions to personal interest, or promotions to career interest, a classifier could use two different binary classifiers, one that identifies personal vs official and another that identifies promotions vs direct messages, and find the intersections of their outputs to classify into one of the four classes expected.

The graph in Fig. 5.1 illustrates a decision boundary of a binary classifier. An optimal decision boundary is the one that maximizes the distance from points in both the classes of data. The solid line represents optimal decision boundary that classifies the two sets. The dotted lines also represent decision boundaries that satisfy the classification requirement of the sets. However they are not optimal boundary lines for the data represented in the Fig. 5.1.

Figure 5.2 represents multi-class classification with four classes of data. The classification here is achieved by using two independent binary classifiers that consider different input features. The classifier has vertical and horizontal decision boundaries. The decision boundaries presented in Figs. 5.1 and 5.2 are linear boundaries that can be achieved through linear regression.

5.1.1 Nonlinear Decision Boundary

Nonlinear decision boundaries can be handled by algorithms such as *support vector machines* (SVMs). Support vector machines are a supervised learning algorithm used for solving binary classification and regression problems. The main idea of support vector machines is to construct a hyperplane such that the margin of separation between the two classes is maximized. In this algorithm each of the data points is plotted as a data point in *n*-dimensional hyperspace. Then construct a hyperplane that maximizes the separation between two classes.

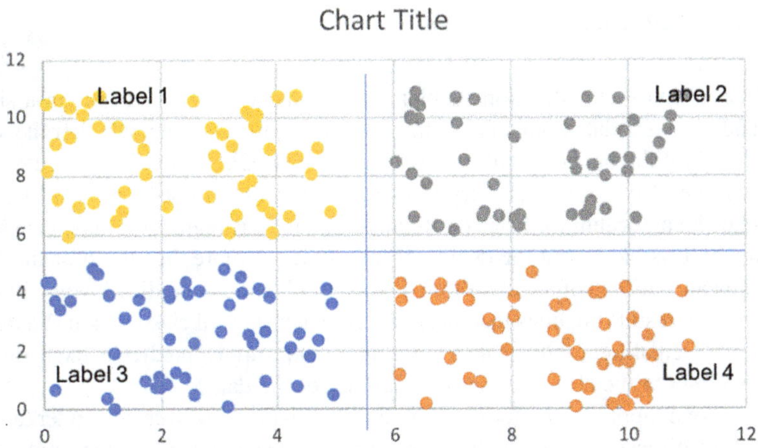

Fig. 5.2 Decision boundary using two binary classifiers

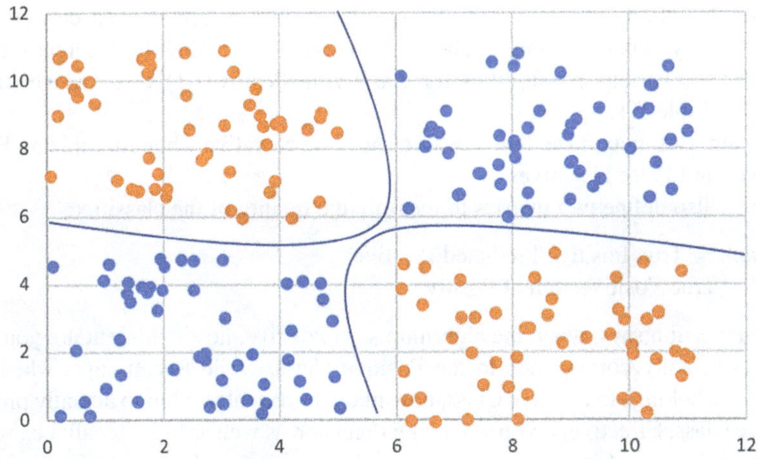

Fig. 5.3 Nonlinear decision boundary

Figure 5.3 represents data of two classes in colors blue and orange. Clearly the class boundaries are not linearly separable. In such a case, SVM uses a technique where two-dimensional representation can be converted to higher dimension data. In this higher dimensional representation, the class boundaries become linearly separable, and SVM classifier can provide a boundary. For example, the data in Fig. 5.3 is transposed to three-dimensional space, with parameters as x, y, and $\sqrt{2}xy$, and then the data in three dimensions is classified using SVM linear plane classifier. The linear plane is then transposed back to two dimensions to form the boundary as shown by the lines in Fig. 5.3. Section 5.4 will talk some more about SVMs.

5.2 Skewed Class

Consider a binary classification problem, where most of the data falls in a single class and very few data points fall in the other class, for example, whether the stock market will hit the upper circuit filter on a given day. Obviously, on most days, the result will be "No" (or 0), and very few days, the result will be "Yes" (or 1). Now, you create an algorithm for Logistic Regression; and this regression may have some inaccuracy. It is likely that the degree of inaccuracy is more than the probability of 1. That means, instead of the algorithm, if you simply always predict 0, you will get a better overall result! So, does that mean that for such skewed classes, you should not bother with creating sophisticated algorithms and simply predict a fixed output, corresponding to the class that has the majority of the data?

The above application does not seem that bad. However, consider a life-threatening disease. Should you always predict that the person does not suffer from the disease, because, anyways, the chances for any given person suffering from life-threatening specific disease are very low? That would be completely wrong!

Thus, for tuning or evaluating of your algorithm for such skewed classes, you cannot simply depend on the number of errors. You need slightly different metric. Consider the various possibilities for classifying Positive (1) vs Negative (0) as shown in Table 5.1.

The aim is to maximize True Positive and True Negative while minimizing False Positives and False Negatives.

Let us also define two metrics to measure the quality of the classifier:

Precision = True Positive/Predicted Positive
Recall = True Positive/Actual Positive

So, Precision says, when the algorithm says Positive, how confident are you that the data actually corresponds to the Positive Class, while Recall says, when the actual data belongs to Positive Class, how likely is your algorithm to actually predict Positive Class. Effectively, you want high Precision as well as high Recall; i.e., when the data is Positive Class, the prediction is also positive, and when the prediction is positive, the actual class is also positive. For example, if you always predict positive, for a data that you know is very likely to be positive, your Recall will be 100%, though Precision would be poorer.

Table 5.1 Possible outcomes against actual results

		Actual observation	
		1	0
Predicted value	1	True Positive	False Positive
	0	False Negative	True Negative

5.2.1 Optimizing Precision vs Recall

In Sect. 3.5.1, you saw that the $h_\theta(x)$ having a value of 0.5 or higher should predict a positive outcome and a lower value can predict a negative outcome. You can change your boundary (i.e., limit), so that instead of 0.5, you take a different value. If you take a higher value, say 0.7, then you will predict Positive Class less often because you narrowed the range of Positive Class to 0.3 (in the range 0.7 and 1). Since the range is closer to 1, when the prediction says Positive Class, there is a very strong likelihood that the actual class is positive. Hence, this algorithm has high Precision. But, Recall value is low, since some of the actually Positive Class may not get predicted as positive. Similarly, if you want to err on the side of caution, you can lower your boundary. This will trigger positive less often.

Figure 5.4 shows a qualitative description of how the change in decision limit effects Precision and Recall. The actual curve will depend on the application and data. So, do not attach numerical significance to the shape of this curve.

The choice of decision point (boundary) that you make will depend on your application and the risk involved. For example, if you are doing a preliminary determination of a life-threatening disease, you should err on the side of the disease being there and, hence, send for secondary, more in-depth opinion.

5.2.2 Single Metric

As you tune your algorithm, you want to be able to use a single metric to compare between two algorithms. You can use a single metric to decide if a specific algorithm (or choice of decision point) is better than the other one. However, in the previous section, you saw the usage of two metrics.

A common single metric that can be considered is *F-score*, whose definition is given in Eq. 5.1:

Fig. 5.4 Precision vs Recall trade-off

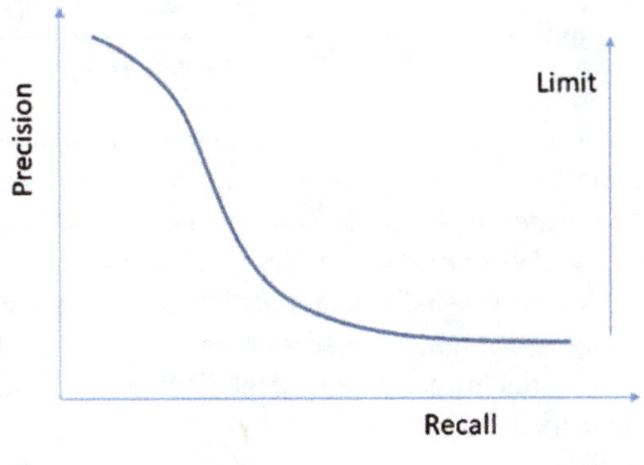

$$F\text{-score} = \frac{2 * \text{Precision} * \text{Recall}}{\text{Precision} + \text{Recall}} \tag{5.1}$$

A common intuition is to use average of Precision and Recall as a common metric. However, F-score is a better metric. If either Precision or Recall is 0, the F-score will be 0, due to the numerator. Thus, this metric prevents you from choosing an algorithm that will always output a fixed value.

You can use the same F-score metric, if you want to decide on your choice of decision point for $h_\theta(x)$.

5.3 Naïve Bayes' Algorithm

In Chaps. 3 and 4, you learnt about Logistic Regression and its application for classification. Naïve Bayes' algorithm is another very simplistic algorithm for classification. Despite its simplicity, it has demonstrated some very good results in classification. The basis for this algorithm is Bayes' theorem, which is stated by Eq. 5.2:

$$P(c|x) = P(x|c) \cdot P(c)/P(x) \tag{5.2}$$

where

$P(c \mid x)$ means probability of c, when x has already been observed
$P(x \mid c)$ means probability of x, when c has happened
$P(c)$ and $P(x)$ denote the probability of c and x, respectively

Consider a situation, where you want to classify a fruit into *apple* and *non-apple*. For ease in understanding, assume that you will do it based on only a single variable – color – and you consider only two colors: *red* and *non-red*. So, given a new object, which you find is *red* in color, the probability of it being an *apple* is given by Eq. 5.3:

$$P(\text{apple}|\text{red}) = \frac{P(\text{apple}) \cdot P(\text{red}|\text{apple})}{P(\text{apple}) \cdot P(\text{red}|\text{apple}) + P(\text{non-apple}) \cdot P(\text{red}|\text{non-apple})} \tag{5.3}$$

During learning stage, you look at all training data, and determine P(apple) as *number of apples/number of total objects considered*. P(red | apple) is determined as the *number of red-apples/total number of apples*. P(red | non-apple) is determined as the *number of red non-apples/total number of non-apples*.

However, usually, classification is rarely done on the basis of single variable. In reality you might have several variables. Naïve Bayes' algorithm assumes that all these variables are independent. This oversimplification explains the *naive* part of the name.

Say, you want to consider weight also. Now, for a given color *red*, and a given weight *W*, the probability of this object being an *apple* is given by Eq. 5.4:

$$P(\text{apple}|\text{red}, W) = \frac{P(\text{apple}) \cdot P(\text{red}|\text{apple}) \cdot P(\text{W}|\text{apple})}{P(\text{apple}) \cdot P(\text{red}|\text{apple}) \cdot P(W|\text{apple}) + P(\text{non-apple}) \cdot P(\text{red}|\text{non-apple}) \cdot P(W|\text{non-apple})} \tag{5.4}$$

Notice in the numerator that the probability of *red* and *W* has been considered independently and multiplied because of assumed independence of the two features. Similarly, in the denominator also, probabilities have been multiplied for *red* and *W*.

Another thing to understand is *weight* is a continuous variable. Hence, during training, you compute the mean and standard deviation of weights of apples and non-apples. That will allow you to determine the $P(W \mid \text{apple})$ (i.e., probability of an *apple* having the weight *W*) and $P(W \mid \text{non-apple})$.

5.4 Support Vector Machines

Support vector machines (SVM) was introduced in the early 1990s and has been successful in applying to real-world classification and regression problems. One of the advantages of SVM is that it can operate with sparse data and the models are generated with relatively small sample set (i.e., training examples).

SVMs provide a compromise between *parametric* and *nonparametric* approaches. *Parametric* approaches use parameters to model, similar to linear regression, whereas *nonparametric* models such as decision trees directly include training data and do not depend on parameters. SVMs are binary linear *classifiers*, i.e., they come up with a hyperplane boundary to separate data points into two classes. They can only handle data points for which there exists a hyperplane boundary. Such data points are called *linearly separable*. SVMs can handle data points that are not linearly separable by mapping data points into higher dimensional space using functions with special properties called *kernels*.

SVM classifier creates a hyperplane of $N - 1$ dimensions for *n*-dimensional feature vectors (for $N = 2$, the hyperplane is a line) to separate data into two classes. The classifier line can be represented by Eq. 5.4:

$$y = w \cdot f(x) + b \tag{5.4}$$

where

$f(x)$ is the feature vector
w is the weight assigned to feature vector
b is the bias term

In a simple case, the feature vector will simply be x, a linear classifier that creates a hyperplane. All values of y greater than $w. f(x) + b$ are classified as class 1, and all other values are classified as class 2. In this linear equation, the feature vector is expected to be linearly representable.

Figure 5.5 is reproduction of Fig. 5.1 with valid class boundaries of a, b, and c represented by red lines. As you can see all of these boundaries classify the data points into two classes correctly. However, line b provides the largest margin for both the classes. SVM looks for boundaries that maximize the margin for the data points using sophisticated *quadratic programming* algorithms (not described in this book). The points closest to the lines a and c represent the *support vectors* that provide the boundary lines for the classes.

For practical problems, there is higher noise in the data, or the data points may not be *linearly separable*. For complex nonlinear boundaries, it can be shown that by converting a nonlinear lower dimensional space into a higher dimensional space, the feature space can become *linearly separable*. Using this theory, classification can be achieved for complex nonlinear boundaries using Eq. 5.4 by simply converting the data to higher dimensional space. For example, Fig. 5.3 shows data points that cannot be classified with linear classifier in two-dimensional space. For these data points, using quadratic terms in a function to represent the data in alternate dimension allows a linear classifier to draw a linear boundary.

Converting data points (i.e., input feature vectors) to higher dimensional space requires mapping them to a function $\phi(x)$, where the function $\phi(x)$ has higher number of variables or higher order variables that can represent the feature variables in the input feature vector. This requires computing distance or similarity measure between each pair of data points using dot product. This can be very computationally intensive since you have to compute $O(N^2)$ dot products for N data points. Moreover, finding the right mapping function can be tricky.

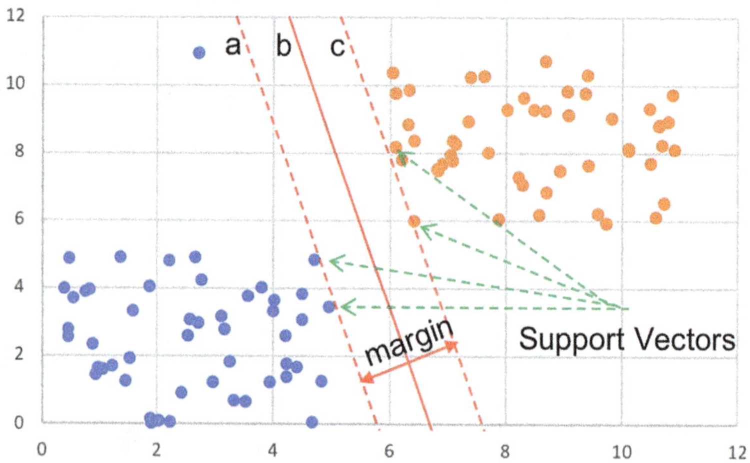

Fig. 5.5 Support vectors and margin representation

To solve this problem, a trick based on Mercer's theorem is used. A simplified explanation of Mercer's theorem states that a positive definite *kernel function* can be decomposed to a dot product. Conversely, instead of computing a mapping function, one can simply use a *kernel function* to represent dot product of mapping function. Hence, simply by choosing a kernel function, computing similarity between data points for a high dimensional feature space is very efficient. This allows computing distance measure without actually computing $\phi(x)$.

Kernels are similarity functions or distance functions that support certain dot product properties. *Kernels* allow substitution of a single function for a higher dimensional feature vector. This kernel function would fit the data model for the input sample space as if it is based on the high dimensional feature vector. This makes computing distance or similarity very easy and also allows constructing the model with small input space. *Kernel* function K can be represented by Eq. 5.5:

$$K(x, y) = \langle f(x), f(y) \rangle \tag{5.5}$$

where

x, y are n-dimensional inputs
$f(x)$, $f(y)$ are functions mapping n dimension to m dimension space
<*a*, *b*> symbol represents a dot product of two vectors a and b

Here is an example to understand the concept of *Kernel* functions better. Say $x = (x1, x2, x3)$, a *three*-dimensional vector. Let function $f(x)$ is defined as in Eq. 5.6:

$$f(x) = (x1.x1, x1.x2, x1.x3, x2.x1, x2.x2, x2.x3, x3.x1, x3.x2, x3.x3) \tag{5.6}$$

where dot represents multiplication of two scalars.

As defined in Eq. 5.6, the function $f(x)$ is a *nine*-dimensional vector. Let y be $(y1, y2, y3)$, and then $f(y)$ can be represented as Eq. 5.7:

$$f(y) = (y1.y1, y1.y2, y1.y3, y2.y1, y2.y2, y2.y3, y3.y1, y3.y2, y3.y3) \tag{5.7}$$

$$\begin{aligned} K(x, y) = \langle f(x), (y) \rangle = \big(& x1.x1 \cdot y1.y1 + x1.x2 \cdot y1.y2 \\ &+ x1.x3 + y1.y3 + x2.x1 \cdot y2.y1 + x2.x2 \cdot y2.y2 \\ &+ x2.x3 \cdot y2.y3 + x3.x1 \cdot y3.y1 \\ &+ x3.x2 \cdot y3.y2 + x3.x3 \cdot y3.y3\big) = (x1y1 + x2y2 + x3y3)\hat{}2 \end{aligned} \tag{5.8}$$

From Eq. 5.8, by using the kernel function that simplifies the dot product, the number of dimensions is reduced from 9 to 3, and the computation for the function is much faster in the algorithm.

Support vector machines (SVMs) use regularization when the data is not completely separable. SVM uses loss function to minimize the empirical error as shown in Eq. 5.9:

$$\text{Error} = \sum\nolimits_{i=1}^{n} L(y_i + f(x_i)) \tag{5.9}$$

There can be infinite number of solutions solving for this equation. SVM constraints the error function by defining the loss as given by Eq. 5.10:

$$\text{Error} = \sum\nolimits_{i=1}^{n} \max\ (1 - y_i f(x_i)) \tag{5.10}$$

5.4.1 Kernel Selection

Kernels can be chosen based on the dataset and characteristics of the dataset. Typically, it is not obvious which kernel works best. It is usually beneficial to start with simple kernels and work up to complex kernels. Based on this approach, one would start with a linear kernel, and if the classification is not performing well, then choose a nonlinear kernel.

Radial bias kernel (RBF) is one of the commonly used kernels in SVM. A RBF kernel is represented as in Eq. 5.11:

$$K(x, y) = e^{-(x-y)^2/2\sigma^2} \tag{5.11}$$

Polynomial and sigmoid kernel functions are some of the other commonly used kernels. They are represented in Eqs. 5.12 and 5.13, respectively:

$$K(x, y) = \left(yx^{\mathrm{T}} + 1\right)^d \tag{5.12}$$

$$K(x, y) = \tanh\left(ayx^{\mathrm{T}} + b\right) \tag{5.13}$$

Cross validation is a good way to determine which of the kernels work best for the data. SVM is particularly useful when the input dataset is small and the number of features is relatively high. SVM can learn with small amount of data for creating a decision boundary.

Support vector machines are an efficient and effective way of classification with small amount of data. Naïve Bayes' has been very effective despite its simplicity. There are multiple techniques to classify including *Neural Networks*. In later chapters introduction of *Neural Networks* will show the use of Neural Networks in classification problems.

Chapter 6
Clustering

Clustering refers to grouping of elements that are close to each other. The assumption being as elements in a groups are close to each other, they would have similarity in properties of interest.

In this chapter you will learn about two basic clustering algorithms:

- K-means, an example of *unsupervised learning*
- K-nearest neighbors (KNN), an example of *supervised learning*

6.1 K-Means

K-means algorithm is about segregating input data into K clusters for a predefined K. Each of the data point in the input set is an unlabelled data (hence, the algorithm is considered unsupervised). The interpretation for each of K clusters can be that the mean value for a cluster is representative of all the elements of that cluster. Or each of the K clusters could represent *a* class of input data.

One of the popular usage of this algorithm is for market segmentation. Suppose, you run a manufacturing unit, where each consumer is unique with unique characteristics and choices. However, you being a mass manufacturer, you may not be able to customize each unit per customer cost-effectively. So, you might want to manufacture a few representative models, and each buyer will purchase the model closest to his/her needs. The characteristics of representative models can then be determined using the K-means clustering by knowing the number of models to manufacture to maximize meeting needs of all customers.

For example, each person in the world has a unique body shape. However, apparel manufacturers segment us into a few sizes: small, medium, large, extra large, etc. By determining number of models of t-shirts to manufacture, apparel manufacturers can then determine the optimal sizes for t-shirts to maximize meeting

G. Rebala et al., *An Introduction to Machine Learning*,
https://doi.org/10.1007/978-3-030-15729-6_6

most of the population needs. This is an example of all of us having been grouped into a few clusters.

Some other applications for *K*-means include analysis of social network, astronomy, identification of compute trends, etc.

6.1.1 Basic Algorithm

Suppose you have *m* datasets, represented by $x^1, x^2, \ldots, x^m$. Each of this dataset is composed of features, $x_1, x_2, \ldots, x_n$. Notice that there is no need for x_0. For most part of this chapter, you would not have to worry about individual features.

The aim is to classify these *m* datasets into *K* clusters and to identify a point for each cluster that represents that cluster (and hence, represent all points within that cluster).

Randomly, identify *K* points, and assume these *K* points to represent the centers of the *K* clusters of your interest.

Now, take each of your data points, and identify which of the *K* centers is closest for each such point, and assign the point to that cluster. At the end of this step, each of your points is assigned to a cluster.

Now, for each cluster, consider all the points that are assigned to this cluster, and find their geometric center. This center is the new representative point for this cluster. At the end of this step, you will have new center for each cluster.

Using this new set of centers, reclassify each of your points once again. Keep iterating over assigning each data point to individual clusters and updating the center of each cluster.

6.1.2 Distance Calculation

The algorithm depends on identifying closest center, which means you need to calculate the distance of a point from each center. The distance computation is given by Eq. 6.1:

$$\text{Distance} = \left\|x^i - \mu_k\right\|^2 \tag{6.1}$$

where

distance is measured between *i*th data and center corresponding to *k*th cluster
μ_k represents center point corresponding to *k*th cluster

Notice that the expression on the right-hand side is actually a square of distance. However, since you are interested in the "closest," rather than the actual distance, you can save compute effort by not performing the square root.

6.1.3 Algorithm Pseudo Code

Let each of the K clusters be represented by C_1, C_2, . . . through C_k. Thus, $\mu_1, \mu_2, \ldots,$ μ_k represent their respective center points.

```
Randomly initialize μ1, μ2, . . ., μk // i.e. K center points representing each cluster
Iterate till no point changes its cluster{
   For each data-point (1 to m)
      Assign it to the Cluster, whose center is closest to the point
   End For
   For each cluster (1 to K)
      Find the mean of all the data-points associated with this cluster
      Update the cluster center(μ) to this new mean found
   End For
}
```

6.1.4 Cost Function

Cost function is a measure that needs to be minimized to find close to optimal solution. While the previous sections have already explained the algorithm, let us understand the cost function that you are trying to optimize. This will help you appreciate some additional fine-tuning to the algorithm.

The main inputs to your algorithm are:

- The number of clusters that you want (i.e., K)
- The training set data: $x^1, x^2, \ldots, x^m$

And the outcome of the algorithm is:

- K clusters, as represented by their means: $\mu_1, \mu_2, \ldots, \mu_k$
- Assignment of each data ($x^1, x^2, \ldots, x^m$), into correct cluster

Let C^i represent the cluster index (between 1 and K) to which data x^i is associated. So, if data x^3 is associated to seventh cluster, C^3 will be 7.

The cost function of interest is to find K clusters with means $\mu_1, \mu_2, \ldots, \mu_k$ and assign each point $x^1, x^2, \ldots, x^m$ into clusters, such that cumulative distance for each point with respect to the assigned cluster's center is minimal. This cost function is given by Eq. 6.2:

$$J\left(C^1, C^2, \ldots, C^m, \mu_1, \mu_2, \ldots, \mu_k\right) = \frac{1}{m} \sum_{i=1}^{m} \left\|x^i - \mu_{C^i}\right\|^2 \tag{6.2}$$

where

$C^1, C^2, \ldots, C^m$ represent the cluster index of each data point x^i

$\mu_1, \mu_2, \ldots, \mu_k$ represent the centers for the clusters
μ_{C^i} represents the center for the cluster to which x^i is assigned

And this cost function is minimized by iteratively updating $C^1, C^2, \ldots, C^m$ and $\mu_1, \mu_2, \ldots, \mu_k$. In Sect. 6.1.3, the first *for-loop* updates $C^1, C^2, \ldots, C^m$ and the second *for-loop* updates $\mu_1, \mu_2, \ldots, \mu_k$, thus reducing the cost function in each iteration until it stabilizes.

6.1.5 Choice of Initial Random Centers

The problem with the algorithm explained so far is that sometimes it may settle at local minima. To overcome this problem, it is better to try with different starting points and find the clusters. For each choice of starting points, apply *K*-means algorithm as explained in Sect. 6.1.3, and compute the cost function of the settled values as given in Eq. 6.2. Finally, use the solution, based on the choice of initial random values that settled at the lowest cost value. Corresponding pseudo code is given as:

```
Iterate { // say over 100 times
    Randomly initialize K centers // μ1, μ2, ..., μk
    Apply K-means algorithm // as explained in section 6.1.3
    Compute cost function // as explained in equation 6.2
}
Now, choose the clustering which has the lowest cost function
```

Usually, if the number of clusters is high (say more than 10), then the problem of local minima does not arise. Hence, the outermost iteration of trying with different combinations of random initialization will not be required.

Another problem that can happen is some of the clusters may not have any point associated to it. This can happen if one of the initial random values chosen is too far off from any of the data points. In Fig. 6.1, points represented by x are actual data

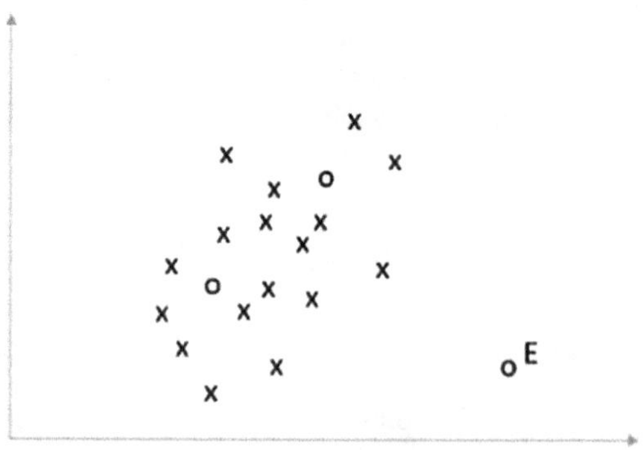

Fig. 6.1 Choice of initial point too far away resulting in empty cluster

points. Points represented by circles are choice of initial random centers. Clearly, the initial choice represented by E will not have any data assigned to it.

In such cases, you can discard the empty cluster. Or if you really need the specific number of clusters, assign your initial random centers to actual data points. This will ensure that the initial value chosen is not too far off.

6.1.6 Number of Clusters

The only remaining decision on the K-means algorithm is how many clusters should you create. One mechanism for such decisions is to use *Elbow method*, which depends on identifying an elbow on the K–J curve. Try different values of K (the number of clusters). For each value, compute the cost function value (represented by J in Eq. 6.2). Plot the curve J as a function of K, as shown in Fig. 6.2.

An *elbow* is visible on the curve. This represents a good choice on the number of clusters. Increasing the number of clusters beyond this does not give any significant reduction in overall cost function value.

Sometimes, the K–J curve may be very smooth, as shown in Fig. 6.3. In such cases, this curve is not very useful in determining the number of clusters.

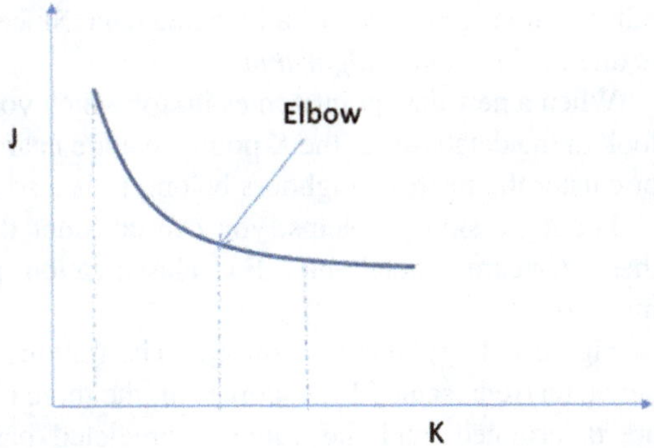

Fig. 6.2 Finding the elbow for determining the number of clusters

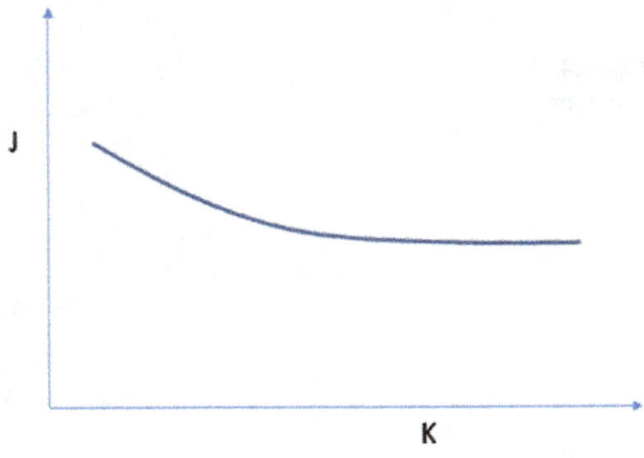

Fig. 6.3 A smooth K–J curve; no elbow visible

In Sect. 6.1.5, you studied the need for trying different combinations of random initialization, in order to avoid local minima. For the purpose of finding the *elbow*, you don't have to worry about trying those iterations. If you hit a local minima, it will be visible in the curve itself. If *J* rises with an increase in *K*, that is indicative of a local minima. Only for that combination, you might want to try another iteration with a different combination of random initialization.

In many cases, the number of clusters is decided based on the end application. For example, an apparel manufacturer might decide to create four clusters, corresponding to sizes (S, M, L, XL); or another manufacturer might decide to create one more cluster – corresponding to XXL.

6.2 *K*-Nearest Neighbor (KNN)

KNN algorithm can be used for both regression problems as well as classification problems. This is a very simple concept of identifying *K*-nearest neighbors for a given data point. The assumption is that similar items are closer together. Here the idea of close is a distance measure that can be a simple Euclidean distance between two points. By finding the largest class of items that are close to the test data, you can conclude that test data belongs to that class. There is actually no learning involved, other than keeping track of all labeled data. Since there is no learning involved, this is often called a *lazy algorithm*.

When a new data point comes in, for which you need to predict the classification, look in the database for the *K* points that are nearest to the new data point. The class or cluster the nearest neighbors belong to is also assigned to the new data point.

For regression problems, you can consider the average of the label values for these *K*-nearest neighbors. For classification problems, you can consider the majority.

Figure 6.4 explains the concept. The training data is represented by *x*. When a new data (represented by *o*) comes in, the three (*K*) nearest neighbors (*A*, *B*, and *C*) are determined, and the value is predicted based on the labels of these three neighbors.

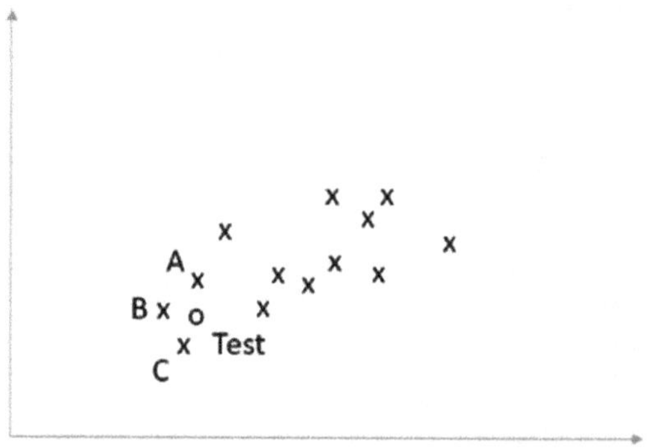

Fig. 6.4 *K*-nearest neighbor

Clearly, predicting the outcome based on the nearest single neighbor ($K = 1$) is a special case. Often, a higher value of K is chosen, in order to not getting affected by a single erroneous data/label. In the example shown in Fig. 6.4, assume B was labeled incorrectly. However, since you are considering three neighbors (A, B, and C), the error due to B will be totally nullified in classification problems and will be diminished in regression problems.

6.2.1 Weight Consideration

You can also decide to assign weight to the contribution from each neighbor. The closer neighbor will have higher weight; and so, assign a weight corresponding to the inverse of the distance from the data to be predicted. This technique is interesting when there are two or more different classes among K-nearest neighbors. Assigning weights will give importance to closeness of neighbors as opposed to just the count of neighbors. Considering the example of Fig. 6.4, point B will have the highest weight. The idea is points which are closer will be more similar to the test data.

6.2.2 Feature Scaling

Recall that the data points can be represented by multiple features. When computing distance between the neighbors, you can use multiple measures like *Euclidean distance* as in Eq. 6.1 or other measures like *Mahalanobis* distance. In these distance measures, the scale of a feature can affect the distance measure disproportionately. Remember to scale the features, as explained in Sect. 4.2. If you don't scale the features, some features will have a higher contribution while calculating the distance.

6.2.3 Limitations

KNN is not very effective for data which have skewed classes. Since the class composition is skewed, the majority decision out of the K neighbors would mostly be in favor of the class which has a majority presence in the whole set of training data.

The other limitation is that the algorithm is sensitive to all dimensions (features) that you capture/consider, even if some feature is totally irrelevant. For example, if you want to predict the stock price, and one of the features that you have been tracking (including for past data), the number of people you saw on the street, the algorithm will consider this feature also – even though, it is irrelevant.

6.2.4 Finding the Nearest Neighbors

As mentioned earlier, this is a lazy algorithm, and there is no real learning involved. The only major work involved is to look into the training dataset and find the *K* neighbors which are nearest to the point of interest.

One obvious method is to compute the distance of this point from all other points of the training data set. And then, find the *K* points of the training dataset, which have minimal distance. Clearly, this is very inefficient method. Over the last several decades, various different algorithms have been proposed which are more efficient, some, even at the cost of slight accuracy.

This is also called post office problem (for $K = 1$), wherein the problem is defined as, for any given address, identify the nearest post-office that should serve this address.

If there are *n* points in the dataset with each data points represented by *d*-dimensions, then computing KNN using obvious method mentioned earlier is performed in *O(nd)* time. The algorithm time can be improved by using *kd*-tree construction and searching within the tree for performance time of *O(log kd).*

The simple idea behind *kd*-tree is that each level of the tree is compared with one dimension. Greater than value is placed on one side of the tree and less than or equal to is placed on the other side of the tree. The data points in the set are chosen at random or in sequence and are placed in the tree on either left or right side. The levels of the tree cycle through the dimensions. Figure 6.5 illustrates tree with two-dimensional data with split dimension as *x* or *y* alternating at different levels of the tree.

To illustrate construction of *kd*-tree, consider data points (50,60), (40,70), (25,45), (35,30), (60,40), and (70,50) to construct *kd-tree*. The tree is illustrated in Fig. 6.6.

The first point (50,60) is constructed as root. The next data point (40,70) is split in *x*-dimension, hence placed on the left side as 40 is less than 50. Next data point (25,45) is placed on left of root and left of (40,70) since this level 2 is split on *y*-dimension and 25 is less than 50 at level 1 and 45 is less than 70 at level 2. Data point (35,30) is left of root, left of (40,70) as 30 is less than 70 and right of (25,45) as 35 is greater than 25 since level 3 is split on *x*-dimension. Data point (60,40) is right of root as 60 is greater than 50 in *x*-dimension comparison. Data point (70,50) is right of root and right of (60,40) as 50 is greater than 40 in *y*-dimension.

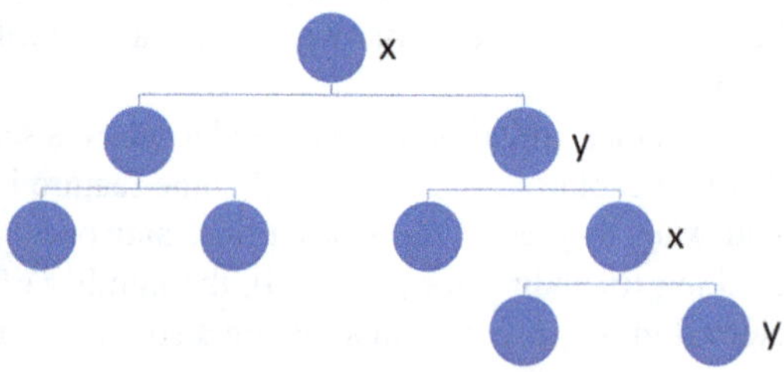

Fig. 6.5 *kd*-tree with two-dimensional data

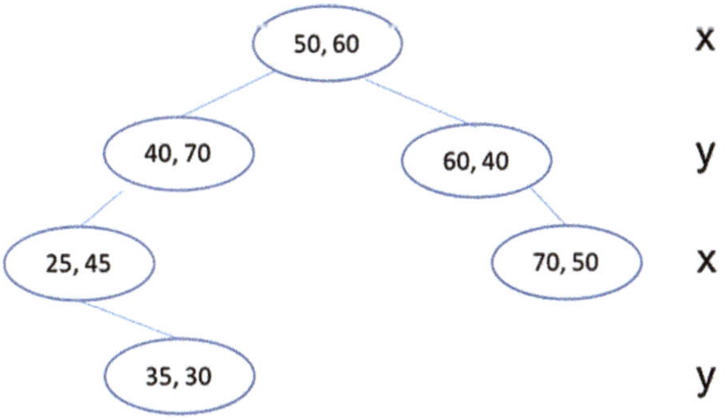

Fig. 6.6 *kd-tree*, for example data

The split space can also be represented in Cartesian coordinates as shown in Fig. 6.7. As a new data point is added to the set, the nearest neighbors are computed based on the tree walk.

The idea behind the tree search is that to find the nearest neighbor, search the whole tree. Remember the closest point found in the tree, to begin with, the node will be parent node. However, prune the subtree to search when the bounding box cannot contain any point closer than closest point already found. Follow the subtree that has the highest chance for pruning.

The algorithm followed in pseudo code is given as:

```
FindNN(point) {
    Start with root node, set as current best
    Recursively go down the tree comparing splitting dimension
        If the splitting dimension is greater for point then go down right subtree
        Else go down left subtree
    At the leaf node, set as current best
    Walk back the tree, for each node traversed on walk back
        If the node is close than current best, update the current best
        Else
            If the splitting dimension is less than current best  for the node then process the other
subtree
            Else continue walk back
}
```

As an example, find a nearest neighbor for a point (45,55). The point will be placed as a leaf node to the tree to the right of point (35,30) similar to the way points are added to tree with splitting on x-dimension and y-dimension in alternative levels. To find the nearest neighbor, using Euclidean distance, the distance to parent is sqrt (725). Set the nearest neighbor as (35,30). As there are no children to the left, compute distance with parent of (35,30), which is (25,45), and the distance is sqrt (500). As (25,45) is closer, the nearest neighbor is updated with (25,45). As there are no left children of (25,45), compute distance with the parent of (25,45) which is (40,70) and the distance is sqrt(250). The nearest neighbor is updated as this distance is lower. Now compute distance with (50,60) which is sqrt(50). As this is the smallest, the nearest neighbor is updated as (50,60). Since point (50,60) is the nearest, the split box on the right should also be searched for nearest neighbor.

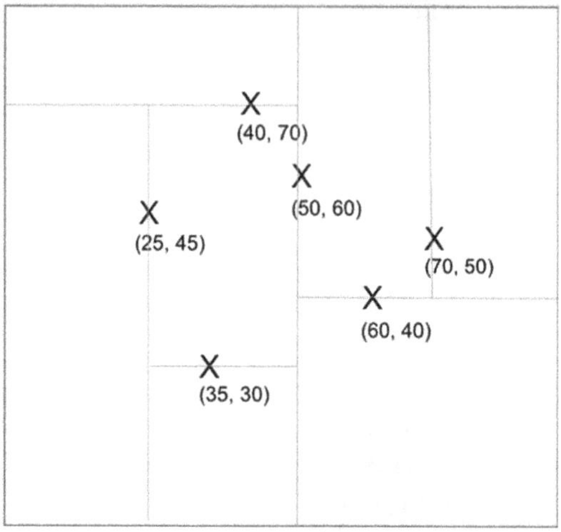

Fig. 6.7 *kd-tree* split space

The point (60,40) is at a distance of sqrt(450) which is higher than the nearest neighbor; hence all the points right of (60,40) need not be considered. Since point (60,40) does not have any left children, the search concludes, and the nearest neighbor of (45,55) is point (50,60).

In the worst case, *NN* may have to search the whole tree, making worst case performance as $O(nd)$ where n is the number of nodes and d is the dimensions. Practically, when using d-dimensions, the order of search is $O(2^d + \log n)$. In summary, the *nearest neighbor* using *kd-tree* is to store partial results so far in search, reduce the subtree to search by comparing splitting dimension, and visit the most promising tree first when not comparing node with splitting dimension. The nearest neighbor algorithm has been successfully applied to recommender systems as it is easy to use and understand with almost no training.

6.3 Next Steps

In this chapter, we introduced K-means clustering, an unsupervised learning algorithm for grouping of data. KNN is a supervised learning algorithm that can classify a new data point into a class very efficiently.

Publicly traded companies are usually grouped under small cap, mid-cap, and large cap companies with their growth dimension as value, core, and growth stocks.

- Using market capitalization and price per earnings growth as metrics run clustering using K-means algorithm on companies traded in Russell 2000 index with 9 clusters. Compare the cluster positioning of your favorite companies with that of the Morningstar rating for those companies. Plot these clusters and see how these boundaries are matching with your expectations.
- Assign labels to the clusters in the companies with their growth and capitalization based on K-means analysis or Morningstar analysis. Select a new company stock that you may be interested in and classify the stock using KNN technique.

Chapter 7
Random Forests

Random Forests are effective and intuitive models used in classification and regression problems. They are intuitive because they provide clear path to a result and are based on underlying *Decision Tree* structures. A *Decision Tree* is a machine learning model built using series of decisions based on variable values to take one path or the other. A *Random Forest* is a collection of Decision Trees that improve the prediction over a single Decision Tree.

Random Forests are supervised machine learning algorithms. As opposed to other machine learning models like neural networks, Random Forests make it easy to see the features that contribute to regression or classification and importance of the variable to the decision.

This section introduces Decision Trees and leads into Random Forests construction and then their usage in classification and regression tasks.

7.1 Decision Tree

A tree is an *acyclic directed* data structure with nodes and edges that connect nodes. A Decision Tree is a tree with nodes representing deterministic decisions based on variables and edges representing path to next node or a *leaf node* based on the decision. Leaf node or terminal node of the tree represents a class label as output of prediction.

For example, given an object, you are asked to classify the object as an apple or an orange or neither, simply by asking questions about the object. Answers to a series of questions regarding the object will lead you to deduce the object classification. Some sample questions you can ask are as follows: is the object edible(?), is the object round(?), is the color of the object Orange(?), can you peel the object skin by hand(?), etc. Each node represents a question and the edge from the node leading to next node represents the path taken based on the response. If the questions are

G. Rebala et al., *An Introduction to Machine Learning*,
https://doi.org/10.1007/978-3-030-15729-6_7

answered correctly, then the last node or terminal node will conclude the class of the object as an orange or an apple or neither.

A *Decision Tree* is modelled on a simple series of questions that lead serially to an answer that best fits the data used in training. Hand-built Decision Trees were commonly used in operations engineering to identify the importance of variable in the decision or to predict the outcome. The questions asked at each decision point or node of a tree lead to a path using "if *a* then *x* else *y*" models.

The decisions made at each node do not need to be binary decisions; however, due to practical reasons, they are usually constructed using binary decisions. To illustrate the increase in number of nodes from binary to ternary split in constructing a Decision Tree, assuming there are 1000 distinct values for a continuous variable, there are 999 possible binary splits and 999*499 or 498,501 ternary splits on a single variable. For simplicity, say values are 1 through 1000 in sequence. Some examples of binary splits are 1 in one group and greater than 1 i.e. 2 through 1000 in the other group or 1, 2 in one group and greater than 2 i.e. 3 through 1000 in the other group and so on forming 999 ways to form binary groups. Some examples of ternary splits are groups with 1 in first group, 2 in second group, and greater than 2 in third group or 1 in first group, 2 and 3 in second group, and greater than 3 in third group and so on forming 999*499 ways of splitting. Due to this explosion in the size of the tree with higher-order splits, binary decisions are most common in constructing a decision tree.

Decision-making in formal or semiformal settings uses variations of Decision Tree structures. For example, consider a store deciding to approve a store credit for purchase of furniture to an individual. Store will start with looking at credit rating of the individual. If credit rating is excellent, then they can approve the credit. If the credit rating is bad, then they reject store credit. If the credit rating is average, then store can check if the person took any new credit in the past 2 months. If there is new credit taken in the past 2 months, then store credit is denied. If there is no new credit issued in the past 2 months and then if the person is a property owner, then store credit is approved. If not a property owner, then the store credit is denied. In real life the decision-making will be different and could be based on many more factors including, income, expenses, outstanding loans, the amount of credit being issued along with the interest rate that can be charged for the credit. The decision representation is shown in Fig. 7.1.

The goal of Decision Tree is to create a model from training data by learning decision rules to predict class or value of target variable. Decision Tree construction is intuitive and can be easily constructed for small number of decision items by hand. For large amount of data, the Decision Tree can be constructed using bagging technique with the rules extracted from the data.

A Decision Tree is built by following three steps:

Step one: Build the root with variables of most importance.
Step two: Build a decision based on the highest information split.
Step three: Recursively construct the nodes and decision using step one and step two until no information can be split on the edge node.

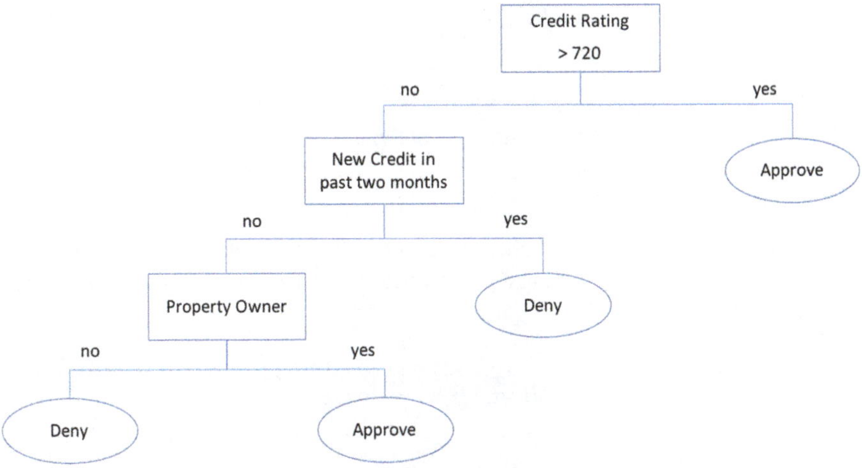

Fig. 7.1 Sample Decision Tree representation

When large number of variables are involved, primary challenge in building a Decision Tree is finding variable or combination of variables of most importance at each of the nodes. This variable selection, also called attribute selection, is generally performed using *information gain* or *Gini impurity* criterion. *Information gain* is used when the variables are categorical, i.e., when the variable values fall into classes or categories and do not have a logical order, for example, types of fruits. *Gini impurity* is used when the variable values are continuous, i.e., the values are numerical, for example, age of a person.

7.2 Information Gain

Using *information theory*, we estimate the amount of information contained within each variable. A key measure in information theory is *entropy*. Entropy quantifies the amount of uncertainty involved in the value of random variable. For example, in binary value outcomes of 0 or 1, if all the outcomes are 1, then the entropy is zero as there is no uncertainty in predicting the outcome. On the other hand, if outcome of 0 or 1 is equal, then the uncertainty or entropy is high.

For a given set of values, if probability of occurrence of a certain value is small, then the information content when the value occurs is high compared to information content of value that is more common or has high probability of occurrence. Information content of random variable X designated as $I(X)$ is computed using $\log(1/p(X))$. Entropy is expectation of information content of all possible outcomes of random variable X. Based on probability mass function of possible outcomes, entropy of random variable X is sum of information content of all possible outcomes of X. Entropy is defined in Eq. 7.1:

$$H = -\sum_{i=1}^{n} (p_i \log p_i) \tag{7.1}$$

where

p_i is probability of occurrence of ith possible value
n is the total number of values
H is the measure of entropy

To understand the entropy better, consider a dice with possible values of 1–6, with each of the outcomes having equal probability of 1/6. The entropy of outcomes of dice is shown in Eq. 7.2:

$$H = -6 * \left(\frac{1}{6}\right) \log_2 \frac{1}{6} \tag{7.2}$$

The entropy calculated from Eq. 7.2 is 2.58. Entropy of a coin flip with equal probability for head or tail can be computed as shown in Eq. 7.3:

$$H = -2 * \left(\frac{1}{2}\right) \log_2 \frac{1}{2} \tag{7.3}$$

The entropy calculated from Eq. 7.3 is 1. From Eqs. 7.2 and 7.3, the entropy for dice outcomes is higher than that of coin toss which is roughly equated to the predictability of the events. Hence, predicting the outcome of dice is more difficult than predicting outcome of coin toss.

When multiple variables are involved, one way to understand the importance of a variable in predicting results is by estimating *information gain* using that variable. Entropy computation for an outcome with and without a variable can be used to understand the *information gain* provided by a variable. The variable with the highest information gain will provide best split for a decision in the Decision Tree. Consider the example in Table 7.1 where *variable 1* and *variable 2* are used to determine whether to continue with the experiment or to stop the experiment.

Information gain is computed by taking the difference in entropy, i.e., entropy conditional on a variable and not conditional on the variable. Entropy of outcomes

Table 7.1 Sample outcomes based on condition of two variables

Variable 1	*Variable 2*	Outcome
3	5	Stop
7	6	Continue
3	3	Stop
4	8	Continue
3	9	Continue
6	5	Stop
5	8	Continue
6	4	Continue

can be computed by identifying number of possibilities of outcomes and their probabilities. There are two possible outcomes: *stop* and *continue*. Probability of outcome *stop* is 3/8, and probability of *continue* is 5/8. Entropy of outcome without controlling for any variable is computed in Eq. 7.4:

$$H(outcome) = -\left(\frac{3}{8}\right)\log_2\frac{3}{8} - \left(\frac{5}{8}\right)\log_2\frac{5}{8} \tag{7.4}$$

Entropy computed from Eq. 7.4 is $H = 0.954$.

The decision to split based on a variable in a Decision Tree can be made based on average value of the variable. Since the average for *variable 1* is slightly more than 4, you would split the Decision Tree based on *variable 1* being greater than 4 or less than/equal to 4.

You can now simplify Table 7.1 into Table 7.2.

Entropy of outcome using *variable 1* alone can be computed by using weighted entropy of each of the branches of split with weights as the probability for each split using Eq. 7.5:

$$H(outcome, variable\ 1) = -p_{(>4)}H_{(>4)} - p_{(\leq 4)}\ H_{(\leq 4)} \tag{7.5}$$

There are four outcomes for *variable 1* greater than 4 and four outcomes with *variable 1* less than 4. The probability of continue is ¾ when *variable 1* is greater than 4 as shown in Table 7.3. The probability of continue is 2/4 when *variable 1* is less than or equal to 4 as shown in Table 7.4.

Entropy computation for outcome with *variable 1* at value >4 is computed in Eq. 7.6, and entropy with *variable 1* ≤4 is computed in Eq. 7.7:

Table 7.2 Decisions based on *variable 1* alone

Variable 1	Continue	Stop
>4 (4 entries)	3 out of 4 times	1 out of 4 times
≤4 (4 entries)	2 out of 4 times	2 out of 4 times

Table 7.3 Decisions based on *variable 1* greater than 4

Variable 1	*Variable 2*	Outcome	Probability
7	6	Continue	3/4
6	4	Continue	
5	8	Continue	
6	5	Stop	1/4

Table 7.4 Decisions based on *variable 1* less than or equal to 4

Variable 1	*Variable 2*	Outcome	Probability
4	8	Continue	2/4
3	9	Continue	
3	5	Stop	2/4
3	3	Stop	

$$H(>4, variable\ 1) = -\left(\frac{3}{4}\right)\log_2\frac{3}{4} - \left(\frac{1}{4}\right)\log_2\frac{1}{4} = 0.81 \tag{7.6}$$

$$H(\leq 4, variable\ 1) = -\left(\frac{2}{4}\right)\log_2\frac{2}{4} - \left(\frac{2}{4}\right)\log_2\frac{2}{4} = 1.0 \tag{7.7}$$

Equation 7.8 shows the evaluation of Eq. 7.5, with respective values weighed from Eqs. 7.6 and 7.7 for a decision split for *variable 1* at value 4:

$$H(outcome, variable\ 1) = \left(\frac{4}{8}\right) * 0.81 + \left(\frac{4}{8}\right) * 1 = 0.9 \tag{7.8}$$

Similarly, the entropy for *variable 2* for the outcome is computed by splitting at >6. Entropy computation is done using Eqs. 7.9, 7.10, and 7.11 – finally evaluation is shown in Eq. 7.12:

$$H(outcome, variable\ 2) = -p_{(>6)}H_{(>6)} - p_{(\leq 6)}\ H_{(\leq 6)} \tag{7.9}$$

For *variable* 2, the probability table based on the variable values is shown in Table 7.5.

Entropy computation for outcome with *variable 2* at value >6 is computed in Eq. 7.10, and entropy with *variable* $2 \leq 6$ is computed in Eq. 7.11:

$$H(>6, variable\ 2) = -\left(\frac{3}{3}\right)\log_2\frac{3}{3} - \left(\frac{0}{3}\right)\log_2\frac{0}{3} = 0 \tag{7.10}$$

$$H(\leq 6, variable\ 2) = -\left(\frac{2}{5}\right)\log_2\frac{2}{5} - \left(\frac{3}{5}\right)\log_2\frac{3}{5} = 0.97 \tag{7.11}$$

Equation 7.12 shows the evaluation of Eq. 7.9, with respective values weighed from Eqs. 7.10 and 7.11 for a decision split for *variable* 2 at value 6:

$$H(outcome, variable\ 2) = \left(\frac{3}{8}\right) * 0 + \left(\frac{5}{8}\right) * 0.97 = 0.28 \tag{7.12}$$

Information gain can be computed for *variable 1* and *variable 2* as shown in Eqs. 7.13 and 7.14, respectively, with *H*(*outcome*) computed in Eq. 7.4:

Table 7.5 Decisions based on *variable 2* alone

Variable 2	Continue	Stop
>6 (3 entries)	3 out of 3 times	0 times
≤6 (5 entries)	2 out of 5 times	3 out of 5 times

$$IG = H(outcome) - H(outcome, variable\ 1) = 0.954 - 0.9 = 0.054 \quad (7.13)$$

$$IG = H(outcome) - H(outcome, variable\ 2) = 0.954 - 0.28 = 0.674 \quad (7.14)$$

Since *variable 2* provides a higher value of information gain, the root of the Decision Tree is based on the *variable 2* with split at value >6. The leaf node will have entropy of 0 as there is no more information gain. Since entropy for split >6 is 0, there are no more decisions to make in that branch of the tree. The decision split is represented in Fig. 7.2.

The decision path for tree with *variable 2* ≤6 has non-zero entropy, and you need to make further splits to build decision tree (Table 7.6). Following the same logic to

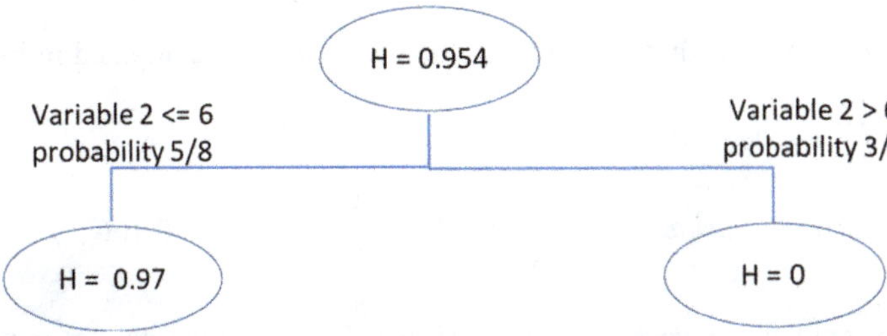

Fig. 7.2 Entropy estimates with probabilities of decision on *variable 2*

Table 7.6 Decisions based on *variable 2* greater than 6

Variable 1	*Variable 2*	Outcome	Probability
4	8	Continue	1
3	9	Continue	
5	8	Continue	

Table 7.7 Decisions based on *variable 2* less than or equal to 6

Variable 1	*Variable 2*	Outcome	Probability
6	4	Continue	2/5
7	6	Continue	
3	3	Stop	3/5
3	5	Stop	
6	5	Stop	

Table 7.8 Decisions based on *variable 1* alone

Variable 1	Continue	Stop
>5 (3 entries)	2 out of 3 times	1 out of 3 times
≤5 (2 entries)	0 out of 2 times	2 out of 2 times

Table 7.9 Decisions based on *variable 1* less than or equal to 5

Variable 1	*Variable 2*	Outcome	Probability
3	3	Stop	1
3	5	Stop	

split, you have to compute information gain for the data using *variable 1* or *variable 2* based on data from Table 7.7.

Decision split using *variable 1* will be at value >5 since we are considering average value of *variable 1* data as split value. The probability of outcome is presented in Table 7.8. The probability distribution is shown in Tables 7.9 and 7.10.

Entropy computation for outcome with *variable 1* at value ≤5 is computed in Eq. 7.15, and entropy with *variable 1* >5 is computed in Eq. 7.16:

$$H(\leq 5, \textit{variable } 1) = -\left(\frac{2}{2}\right)\log_2\frac{2}{2} - \left(\frac{0}{2}\right)\log_2\frac{0}{2} = 0 \tag{7.15}$$

$$H(> 5, \textit{variable } 1) = -\left(\frac{2}{3}\right)\log_2\frac{2}{3} - \left(\frac{1}{3}\right)\log_2\frac{1}{3} = 0.276 \tag{7.16}$$

Weighted entropy with decision split using *variable 1* is computed in Eq. 7.17:

$$H(\textit{outcome}, \textit{variable } 1) = -p_{(>5)}H_{(>5)} - p_{(\leq 5)}\,H_{(\leq 5)} \tag{7.17a}$$

$$H(\textit{outcome}, \textit{variable } 1) = \left(\frac{2}{5}\right) * 0 + \left(\frac{3}{5}\right) * 0.276 = 0.166 \tag{7.17b}$$

Decision split using *variable 2* will be at value ≥5 since we are considering average value of *variable 2* data for split value. The probability of outcome is

Table 7.10 Decisions based on *variable 1* greater than 5

Variable 1	*Variable 2*	Outcome	Probability
6	4	Continue	2/3
7	6	Continue	
6	5	Stop	1/3

Table 7.11 Decisions based on *variable 2* alone

Variable 1	Continue	Stop
>5 (3 entries)	1 out of 3 times	2 out of 3 times
≤5 (2 entries)	1 out of 2 times	1 out of 2 times

Table 7.12 Decisions based on *variable 2* less than 5

Variable 1	*Variable 2*	Outcome	Probability
3	3	Stop	1/2
6	4	Continue	1/2

Table 7.13 Decisions based on *variable 2* greater than or equal to 5

Variable 1	*Variable 2*	Outcome	Probability
3	5	Stop	2/3
6	5	Stop	
7	6	Continue	1/3

presented in Table 7.11. The probability distribution is shown in Tables 7.12 and 7.13.

Entropy computation for outcome with *variable 2* at value <5 is computed in Eq. 7.18, and entropy with *variable 1* >5 is computed in Eq. 7.19:

$$H(<5, \mathit{variable}\ 2) = -\left(\frac{1}{2}\right)\log_{2}\frac{1}{2} - \left(\frac{1}{2}\right)\log_{2}\frac{1}{2} = 1 \tag{7.18}$$

$$H(\geq 5, \mathit{variable}\ 2) = -\left(\frac{2}{3}\right)\log_{2}\frac{2}{3} - \left(\frac{1}{3}\right)\log_{2}\frac{1}{3} = 0.276 \tag{7.19}$$

Weighted entropy with decision split using *variable 2* is computed in Eq. 7.20:

$$H(\mathit{outcome}, \mathit{variable}\ 1) = -p_{(\geq 5)}H_{(\geq 5)} - p_{(<5)}\,H_{(<5)} \tag{7.20a}$$

$$H(\mathit{outcome}, \mathit{variable}\ 1) = \left(\frac{2}{5}\right) * 1 + \left(\frac{3}{5}\right) * 0.276 = 0.676 \tag{7.20b}$$

Information gain can be computed for *variable 1* and *variable 2* as shown in Eqs. 7.21 and 7.22, respectively. Here *H(outcome)* is the value computed for left tree in Eq. 7.11:

$$IG = H(\mathit{outcome}) - H(\mathit{outcome}, \mathit{variable}\ 1) = 0.97 - 0.276 = 0.694 \tag{7.21}$$

$$IG = H(\mathit{outcome}) - H(\mathit{outcome}, \mathit{variable}\ 2) = 0.97 - 0.676 = 0.294 \tag{7.22}$$

Since *variable 1* provides a higher value of information gain, the root of the Decision Tree is based on the *variable 1* with split at value ≥5. The decision split is represented in Fig. 7.3.

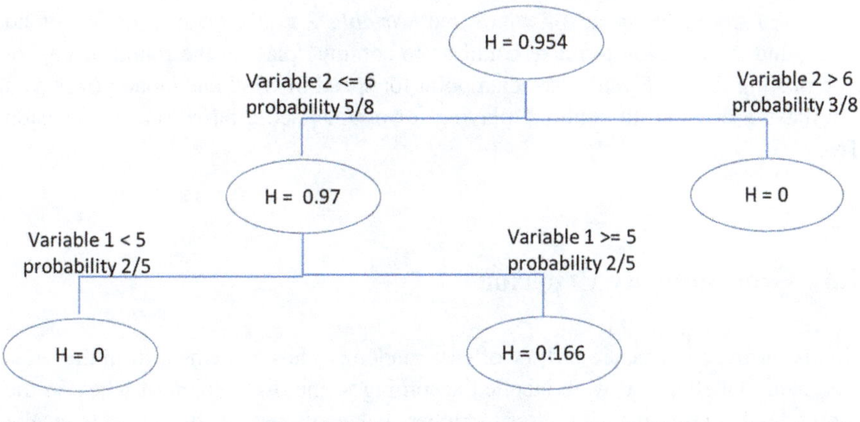

Fig. 7.3 Entropy estimates with probabilities of decision on *variable 2* and then *variable 1*

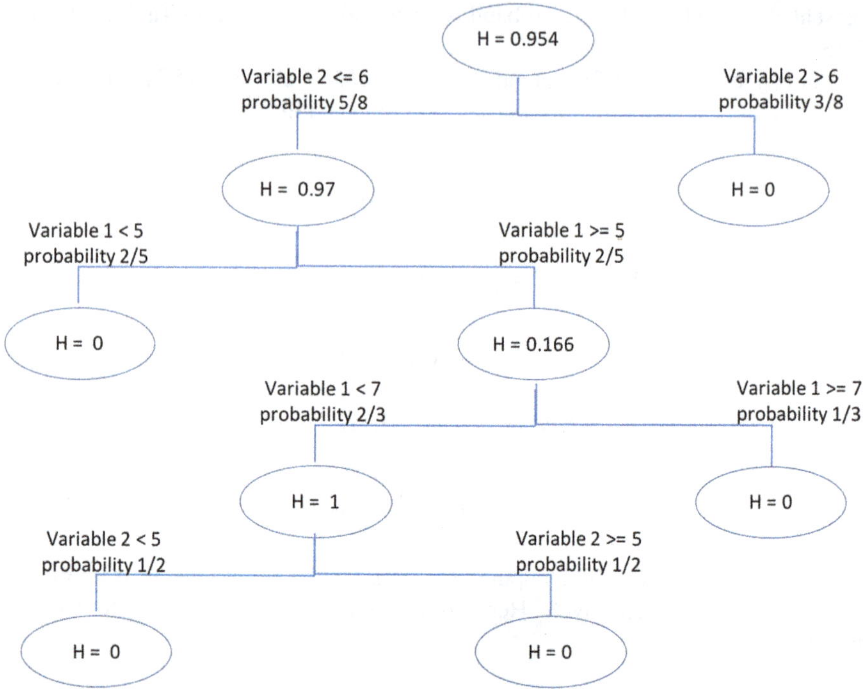

Fig. 7.4 Decision Tree with entropy estimates

Further splitting data on the tree branch with values *variable 1* greater than 5 will result in the decision to split on *variable 1* with the result as shown in Fig. 7.4. This split computation is left to reader as an exercise.

The constructed decision tree is represented in Fig. 7.5. Since the amount of data is relatively small, there is a chance of overfitting. This Decision Tree is ready to make decisions to continue or stop based on the new data point. For example, *variable 1* could be speed of wind, and *variable 2* could be measure of cloud cover, and the decision to make could be to continue playing the round of golf or stop playing. When given a new data point for speed of wind and cloud cover, you can make a decision to continue playing for optimal score based on the Decision Tree.

7.3 Gini Impurity Criterion

Gini impurity is a measure of how often a randomly chosen element from the set is incorrectly labelled if it were labelled according to the distribution of labels in the subset. Higher Gini impurity refers to higher chance of misclassification conversely,

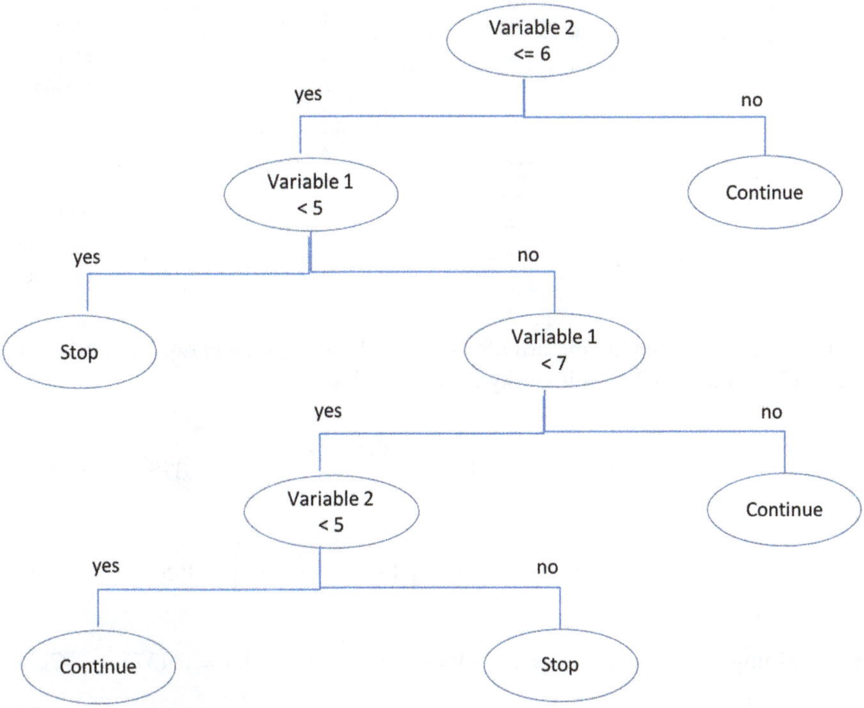

Fig. 7.5 Decision Tree for predicting outcome

and lower impurity refers to lower chance of misclassification. In constructing a Decision Tree split, the goal is to split with the lowest weighted Gini impurity value for the child trees. This translates to split with the largest reduction in weighted Gini impurity for child nodes from Gini impurity of parent node. In other words, the split will result in largest possible homogeneity in the child node. To compute Gini impurity for a set of items with J classes, the computation is represented in Eq. 7.15:

$$Ic(p) = \sum_{i=0}^{J} \left(p_i * \sum_{k \neq i}^{J} p_k\right) = \sum_{i=0}^{J} p_i(1 - p_i) = 1 - \sum_{i=0}^{J} p_i^2 \quad (7.23)$$

where

$Ic(p)$ is the Gini impurity
p_i is the probability of ith label
J is the number of classes in the set

Since the sum of all the probabilities of outcomes or labels is equal to 1, the Gini impurity computation has been simplified in Eq. 7.23 by substituting $\sum_{k \neq i}^{J} p_k$ with $1 - p_i$ as $\sum_{k=1}^{J} p_k$ is 1. Using the same example from previous section, repeating the data for convenience is shown in Table 7.14.

Table 7.14 Table 7.1 reproduced for convenience

Variable 1	*Variable 2*	Outcome
3	5	Stop
7	6	Continue
3	3	Stop
4	8	Continue
3	9	Continue
6	5	Stop
5	8	Continue
6	4	Continue

For *variable 1*, the Gini impurity for split at >4 is computed in Eq. 7.24 and ≤4 in Eq. 7.25, and the final Gini impurity is computed in Eq. 7.26:

$$\text{Gini impurity}(>4) = 1 - \left(\left(\frac{3}{4}\right)^2 + \left(\frac{1}{4}\right)^2\right) = 0.375 \tag{7.24}$$

$$\text{Gini impurity}(\leq 4) = 1 - \left(\left(\frac{2}{4}\right)^2 + \left(\frac{2}{4}\right)^2\right) = 0.5 \tag{7.25}$$

$$\text{Gini impurity}(\mathit{outcome}, \mathit{variable}\ 1) = \frac{4}{8} * 0.375 + \frac{4}{8} * 0.5 = 0.4375 \tag{7.26}$$

For *variable 2*, the Gini impurity for split at >6 is computed in Eqs. 7.27, 7.28, and 7.29:

$$\text{Gini impurity}(>6) = 1 - \left(\left(\frac{3}{3}\right)^2 + \left(\frac{0}{3}\right)^2\right) = 0 \tag{7.27}$$

$$\text{Gini impurity}(\leq 6) = 1 - \left(\left(\frac{2}{5}\right)^2 + \left(\frac{3}{5}\right)^2\right) = 0.48 \tag{7.28}$$

$$\text{Gini impurity}(\mathit{outcome}, \mathit{variable}\ 2) = \frac{3}{8} * 0 + \frac{5}{8} * 0.48 = 0.3 \tag{7.29}$$

From the Eqs. 7.26 and 7.29, the weighted Gini impurity for split on *variable 1* (0.4375) is higher than the weighted Gini impurity for split on *variable 2* (0.3). Hence, the *variable 2* will be chosen as the root variable with the split at >6. In other words, selecting split on *variable 2* provides the largest reduction in Gini impurity for the outcome. This process is continued recursively to build the entire Decision Tree.

For the same data using Gini impurity measure and entropy measure, different variables can selected as root node. This is because of bias of each algorithm. The Gini impurity is biased toward selecting wider spread of data for variable selection, whereas the entropy method is biased toward compact data with lower spread.

7.4 Disadvantages of Decision Trees

In a Decision Tree, the choice is made to optimize the decision at each of the nodes. Choosing best result at each step does not result in global optimal result. Usually, greedy algorithms are used at each step to determine the decision split at each node.

Another drawback with Decision Trees is that they are prone to overfitting, i.e., generalizing decision or force fitting a decision based on one data point. If the tree is deep, the number of samples considered at each decision becomes small. Consider predicting a winner in game of football between San Francisco and Seattle. The decision can be based on the location of the game being played for home team advantage, past results of games between two teams, if the best players on the teams are playing on the given day, the results of previous game for each of the teams with other teams, weather conditions on the given day, etc. As the number of factors being considered increases, the data points available for the combinations are not significant to assign sensible probability for outcome. Hence, the decision may be based on one available data point which maybe an anomalous data point causing overfitting for the data.

7.5 Random Forests

Random Forests are collection of decision trees. Random Forests are an ensemble classifier using many Decision Tree models. Random Forests are way of averaging multiple Decision Trees built from different parts of the training set with the goal of reducing overfitting by a single Decision Tree. The increased performance comes at a cost of some loss of interpretable and accurate fit to training data.

Random Forests use *bagging* to construct multiple Decision Trees. Bagging or *bootstrap aggregation* is a technique of sampling a subset of training data with replacement and constructs the model based on the sampled training set. For a large dataset D, it is expected to have ~62.3% of unique samples from D in each of the subsets.

Random Forests grow many Decision Trees by sampling from the input dataset D for a set of features instead of the training data samples. The process is also called *feature bagging*. If one of the features is highly correlated, then that feature will be selected by many different trees. This causes the trees to be correlated and thus increases the bias of the algorithm, reducing the accuracy. Typically, for a classification problem with n features originally, $\sqrt{n}$ features are used in each split. For regression problems, number of features recommended is $n/3$.

7.5.1 Data Bagging

Consider *bagging* as putting original dataset into different bags. Data is randomly chosen from the original D dataset and puts into various bags. If you have to create B bags, for each bag, randomly choose some (actually ~62.3%) data from the original dataset D, and put into the bag. In this way, put the data in all the bags. Since this sampling of data is with replacement, the same data can be present in multiple bags. The same data can also potentially go multiple times in the same bag just that duplicate data in the same bag is effectively useless in that bag, from the perspective of creating a decision tree for that bag. Each of the bag so created will form one tree. Usually, the tree also utilizes *feature bagging* described in Sect. 7.5.2.

7.5.2 Feature Bagging

Each of the B bags so formed still has n features. For each tree, you should create feature bagging, i.e., randomly chose some ($\sqrt{n}$ or $n/3$) features, and you will use only those features for creating a decision tree for the data in that bag. For any data bag, create multiple sets of feature bags. Use *cross validation* to decide which feature bag is most apt for this specific data bag. And use this feature bag for the data to create a Decision Tree.

So, each data bag will have a different set of features chosen for creating the Decision Tree.

7.5.3 Cross Validation in Random Forests

In Random Forests, cross validation is estimated internally during the run because of the bagging procedure.

For each bag, based on sampling, about 62% unique samples from original dataset are used. Hence, about a third of the samples are not present in ith tree construction. These left out data will be used for cross validation, to identify the most useful feature set combination, among the multiple randomly chosen feature bags for that data bag.

The proportion of sample classified incorrectly from the cross validation set (for that data bag) over all classifications of the cross validation set is the error estimate of the system.

This error estimate is also referred to as *out-of-bag* (OOB) error estimate. OOB is mean prediction error on each training sample j, using only the trees that did not have sample j in the bootstrap sample.

7.5.4 Prediction

So, at the end of bagging, you now have *B* Decision Trees, each tree with reduced dataset and reduced feature set.

For a new data point, prediction of its value is aggregation of predicted values from the trees. The prediction can be categorical or regression based on the Decision Tree construction. The final result is aggregation of results from all the Decision Trees as shown in Eq. 7.3:

$$f = \frac{1}{B} \sum_{n=1}^{B} (f_n(x)) \tag{7.30}$$

where

f is the final prediction from the Random Forest
B is the number of trees constructed from dataset
n is the index of Decision Tree constructed
f_n is the result from Decision Tree n
x is the input sample that may not be in the training set, for which prediction is desired

7.6 Variable Importance

When a large number of variables are involved in decision-making, Random Forests can be used to determine importance of the variables involved and order them with respect to their importance in the decision-making. The importance of variables is derived from mean decrease in impurity of each variable and the degree of interaction of each input variable with other input variables described in context of Decision Tree in Sects. 7.2 and 7.3.

Breiman (2001) describes the variable importance computation using the OOB technique discussed in Sect. 7.5.3. The intuitive notion in determining the variable importance is that if the variable is important, then rearranging the values of the variable in constructing the trees will not reduce the prediction accuracy.

In practice, for a variable *m*, compute the number of correct classifications of the tree for out-of-bag cases. Permute values of variable *m* and then compute the classification of out-of-bag case for the tree. Compute the difference in number of correct classifications after permutation and before permutation. The average difference over all the trees is the importance score of the variable *m*.

Determining variable importance is useful in improving the results by constructing trees with subset of variables when large number of variables are involved in decision-making. Variable importance also provides intuitive understanding of importance in applications like genetics and neurosciences.

When multiple correlated variables are present that have significant effect on the outcome, then the variable importance method will have difficulty in identifying relative importance of these variables.

7.7 Proximities

Proximity of two samples is true if they follow the same decision path to the outcome in a tree. *Proximity count* provides a measure of how frequently unique pairs of training samples end in the same terminal node. Proximity values in a tree are computed by incrementing the proximity value for each case in the input set that end in the same terminal node by one. If there are N input cases, then the proximity matrix is of size $N \times N$. In a given tree of the forest, this value is either 0 or 1 for a pair of input observations. Proximity values are obtained by averaging over all trees in the forest and normalizing the proximity value by dividing with the number of trees.

Proximity values can be used as a distance measure between pairs of inputs. The distance measure intuitively can be treated similar to *Mahalanobis* distance or Euclidean distance. This measure can be used in clustering and finding outliers.

Proximities are useful in computing outliers, missing data in the dataset, and computing prototypes. One method to replace missing data is to find all the cases that end in same class as the case with missing data of variable k in the input and then find the median value for variable k that end in class n. This method provides fairly good approximation for missing data in the dataset.

7.7.1 *Outliers*

Outliers can be defined as cases that are different from expected conditions from the model expected from the data. Using proximities, outliers are defined as the cases where the proximity values are small.

Outlier value can be computed by normalizing proximity score across all the samples. Compute raw outlier measure by dividing the total number of samples with average proximity score of an input case. If the proximity score is small, then this value will be high. Compute median of all the raw outlier scores, and normalize the raw outliers based on standard deviation from absolute median and median values. The larger the normalized value, the higher the outlier probability:

$$\text{Raw score} = \sum\nolimits_{\text{class}(k=j)} \text{proximity}(n, k) \tag{7.31}$$

where

j is the class of outcome in a tree
n is the input case that ends in outcome class j

$$\text{Outlier score} = \frac{\text{raw score} - \text{median}}{\sigma} \tag{7.32}$$

where

median is median of all raw scores of all classes of outcomes
σ is standard deviation of all raw scores of all classes of outcomes

As a rule of thumb, input cases with an outlier score of 3 or more from Eq. 7.32 are considered outliers. These values can be better interpreted through visualization of plot of all outlier scores.

7.7.2 Prototypes

Prototypes are useful in identifying how variables are related to classification. For a given class j, find a case (l) that has the largest number of outcomes as class j. Find all the cases that are nearest neighbors (k) of case l using the proximity numbers. Among these k nearest neighbors cases, find 25th percentile and 75th percentile values for each variable. The median values are prototypes for class j. The quartile values provide an estimate of stability of each variable for the class j. The prototype can be thought of as a similar measure of cluster center as representative of cluster in k-means clustering.

7.8 Disadvantages of Random Forests

Random Forests can overfit with the data when there are large number of high cardinality categorical variables. For example, with two input variables that have 100 and 200 distinct values, there are 200,000 unique combinations that could lead to a decision. In training the forest, the data will fit to the data too well and may not perform well with the validation on new data.

Random Forests also provide discrete models for prediction. If the response variable is a continuous variable, then there are only n distinct values that are possible through prediction. Other regression mechanisms would provide a model for continuous data prediction.

Random Forests unlike Decision Trees are not easy to interpret. As the result is from ensemble of trees, interpretation from intuition point of view is very hard.

Random Forests do not support incremental learning very well. With new data they have to be relearned.

Random Forests also tend to be slower than other regression prediction models, particularly, when there are higher depth and large number of trees.

7.9 Next Steps

In this chapter, we introduced Decision Trees for *classification* and *prediction* for outcomes. Decision Trees are prone to *overfitting*, and Random Forests are introduced that improve results of Decision Trees. The following exercise will help you to reinforce the learnings from material from this chapter.

- Try to predict whether the closing price of a stock or index based will be higher or lower for the day based on last 10 days of data. Choose a stock/index of your choice and collect data for metrics like volume, closing price, moving average, beta and any other performance metrics you can identify. Perform the following analysis using data for key performance metrics like volume, closing price, short interest, beta, moving average for 5 days, 1 month, 3 months and other metrics for which data is collected.
  - Use information gain and Gini impurity methods to look for variable of importance and decisions reached by both methods.
  - Use Random Forests feature bagging and data bagging methods. Verify if the results are improved over Decision Tree.
- Try to predict whether the next closing price range will be higher or lower by 0–5%, 5–10%, 10–20%, etc. by adding longer-term data.

Chapter 8
Testing the Algorithm and the Network

The learnings in this chapter will help you determine if your choice of features and the number of datasets are sufficient or should you increase datasets and/or increase/ decrease the number of features forming your hypothesis.

In some of the prior chapters, you saw how to create a hypothesis and assign values to various parameters based on the training data. In the next chapter, you will also learn about *neural networks* for solving similar problems.

The general rule of thumb says that the more datasets you have, to train your algorithm or network, the better will be your final model (or network). However, if you don't have the right set of features, higher number of datasets may not give any additional accuracy. On the other hand, even if you have the right number of features, if you don't have enough datasets, the parameter values will not get the right values for higher accuracy in prediction. This chapter focuses on identifying the conditions of model fit and actions required for improving accuracy.

8.1 Test Set

Section 4.4.3.1 explained the concept of *cross validation set*, which can be used to compare and choose between two sets of hypotheses (or networks) or between two values of λ (the regularization parameter).

Similar to cross validation set, you should also have a *test set*.

Randomly assign each data from your dataset into one of:

- Training set (~60%)
- Cross validation set (~20%)
- Test set (~20%)

The decision on hypothesis should never be taken based on the data in training set. Training set should be used only for finding the parameters (i.e., values) of the hypothesis (or the network). For an algorithm trained on a specific set of data

G. Rebala et al., *An Introduction to Machine Learning*,
https://doi.org/10.1007/978-3-030-15729-6_8

(training data), when evaluated on the same set of data, it will always provide a highly accurate result. What you want is an algorithm that will provide a good accuracy on a new set of data. Hence, algorithm should be evaluated on cross validation data or test data.

Once a hypothesis has got all its parameters defined (based on training set), both *cross validation set* and *test set* are new data for it, and hence, they both are expected to show similar error characteristics.

8.2 Overfit

Section 4.4 explained a concept, where the algorithm made the curve extra complex in order to ensure that the curve could satisfy all of the training data. However, when a new data was presented to this complex curve, it could potentially give an incorrect result.

This is considered a problem of *overfit*, also called as *variance problem*. This typically can be corrected by increasing the value of λ (the regularization parameter).

8.3 Underfit

The exact opposite problem of *overfit* is that of an *underfit*, also called as *bias problem*. In this case, the algorithm has too few features, and hence, it does not account for various relations that it should account for. This can be solved by adding higher-order variables and/or reducing the value of λ.

8.4 Determining the Number of Degrees

Suppose, you identify five variables x_1 through x_5 that are useful in predicting y. x_0 is always added implicitly, with a value of 1. The first thing you want to do is to identify how many degrees do you want to consider.

0 degrees would mean: x_0

1 degree would mean: $x_0, x_1, x_2, x_3, x_4, x_5$

2 degrees would mean: $x_0, x_1, x_2, x_3, x_4, x_5, x_1^2, x_2^2, x_3^2, x_4^2, x_5^2, x_1*x_2, x_1*x_3, x_1*x_4, x_1*x_5, x_2*x_3, x_2*x_4, x_2*x_5, x_3*x_4, x_3*x_5, x_4*x_5$

and *so on.*

In order to determine the right number of degrees that you need for a good algorithm, plot a graph which has:

- The number of degrees on the horizontal axis

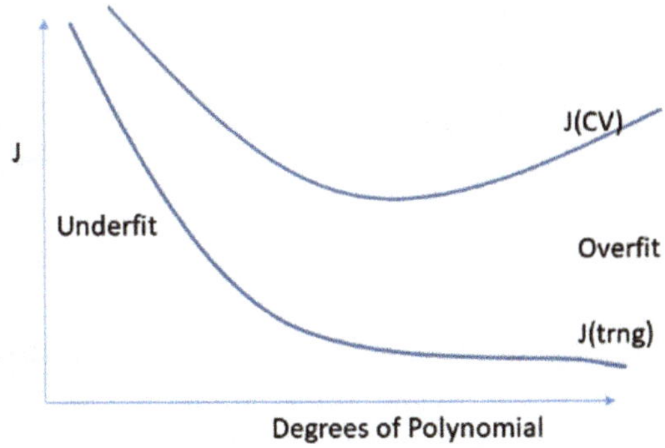

Fig. 8.1 Cost function against polynomial degree

- Cost function for training data (J_{trng}) and cost function for cross validation data (J_{cv}) on the vertical axis.

The general shape of the graph should look similar to Fig. 8.1.

The left side of the graph indicates a case of underfit. The curve obtained is too simple, even the training data cannot fit well. This is similar to a very high value of λ, which effectively retains only the lowest-order variables. Say, in extreme case, only x_0 survives.

The right side of the curve indicates a case of overfit. The training data fits well because the curve has now enough variables to account for all the training data. However, a new set of data (cross validation) does not fit well. The algorithm has brought in features which should not be playing a role in predicting, and these features (along with the associated parameters) have been artificially used to fit known data; however, with new data, these useless features (and parameters) are causing more harm than good.

The correct degree of polynomial to be used corresponds to the point, whenever J (CV) is minimal.

8.5 Determining λ

Once you have determined the degree of the polynomial, you now want to determine λ. A proper choice of λ will weed out unnecessary combinations of variables from the polynomial.

Section 4.4.3 already explained how to choose the right value of λ. Figure 8.2 shows how the graph plot would look for various choices of λ.

The right side of the curve indicates a very high value of λ. That means fewer number of features survive. Thus, this is a case of underfit. With fewer number of variables, even the training data is not able to be fit properly, and you get a higher error – both with training data and cross validation data.

The left side of the curve indicates a very low value of λ. This allows higher degree of flexibility with more parameters, and any known data can be fitted well.

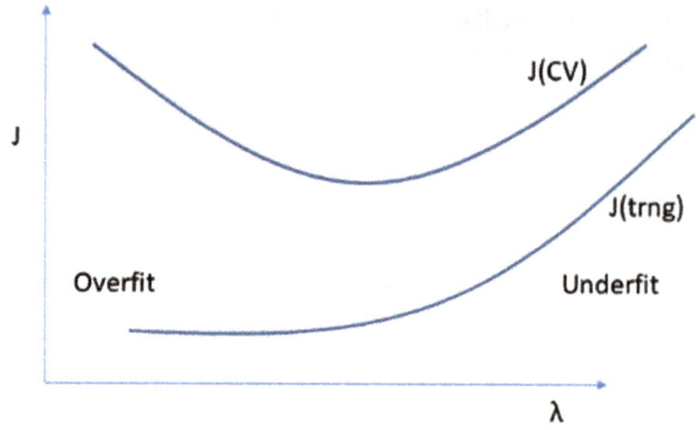

Fig. 8.2 Choice of λ

Thus, training data will show lesser error. However, the same parameters and the values may not be correct for a new data. Thus, cross validation data could have higher error.

8.6 Increasing Data Count

Consider using a very small set of training data. In such a case, it should be possible to identify parameter values such that all the data fit perfectly, resulting in 0 error for the training data. And, as the number of training data increases, it may no longer be possible to fit all the data perfectly. This will be reflected in an increase in error on the training data. The effect on cross validation data will though be different. With higher number of training data, you will get a better value for the parameters, and hence, the cross validation set will show a lower error. Figure 8.3 shows how the cross validation and training errors plot with an increase in the number of training data.

8.6.1 High Bias Case

If there is a *high bias*, the parameters don't fit the test data itself well. And they are not likely to fit the cross validation data as well. Thus, both cross validation set and test set will show a high value of error. Figure 8.4 shows how the error curves plot for *high bias* situations, even if the number of training data were to increase.

You will see that the error values are high for training data, almost since the beginning, and that cross validation data refuses to come down below a certain value.

In such cases, there is no value in adding more training data. The model should be fixed first, to reduce the bias.

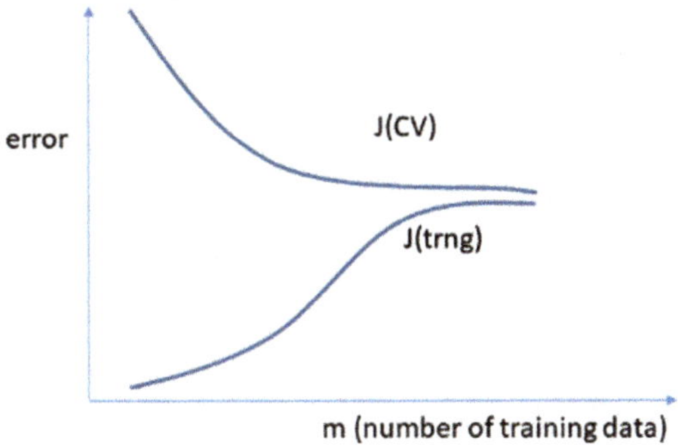

Fig. 8.3 Variation of cross validation and training errors with an increase in training data

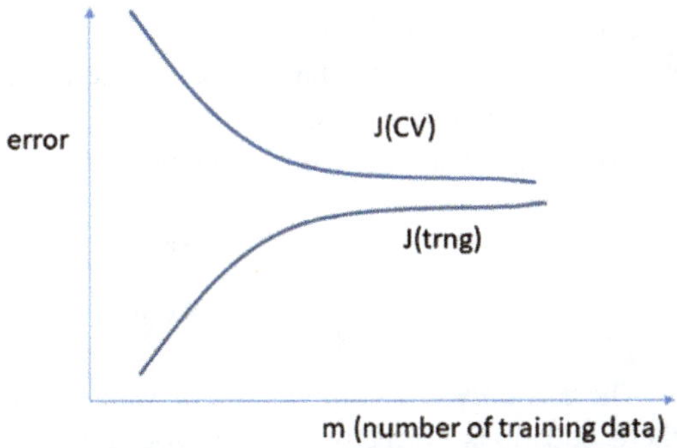

Fig. 8.4 Error curves for high bias case

8.6.2 High Variance Case

For *high variance* also, the error on cross validation set is high; however, this situation differs from the high bias situations because here the error on training data will typically be much lower, due to overfitting. Figure 8.5 shows how the error plots look for high variance situations.

If you extend the curves to the right, the training data error will increase, as not all data can fit a given number of features, and cross validation error will decrease, as higher number of training data provides better parameter values. That implies that for overfitted case, an increase in the number of training data can help improve the model accuracy.

8.7 The Underlying Mathematics (Optional)

Suppose you have to find the values of some variables and you need to find those values by solving some simultaneous equations. The parameter values are the variables that you want to find, and the training data form the equations to solve – to find the parameters.

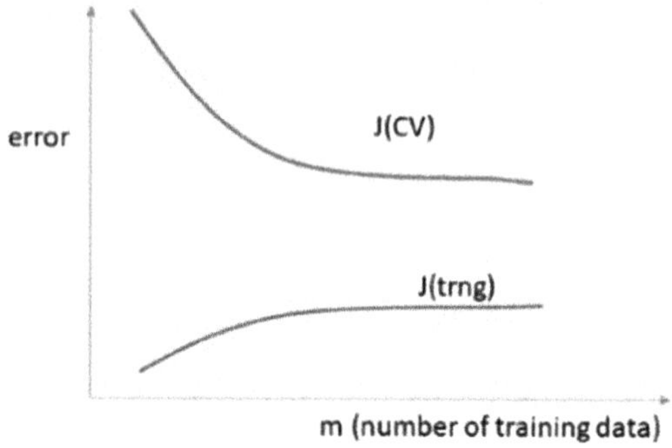

Fig. 8.5 Error curves for high variance vase

Consider that you have n number of variables, whose value you want to determine. Basic algebra says that you need exactly n number of independent equations to solve for these n variables.

Consider the simultaneous equations:

$$3x + 5y = 13 \tag{8.1}$$

$$4x + 8y = 20 \tag{8.2}$$

There is only one combination of values ($x = 1$; $y = 2$) which will satisfy both these equations correctly. Hence, you got an exact solution for n variables and n equations.

If you have less than n equations, you will get many possible solutions. Assume you have only one equation, viz., Eq. 8.1. You can get many (actually, infinite) possible values. Some combinations of values are:

$x = 0$, $y = 2.6$
$x = 1$, $y = 2$
$x = 2$, $y = 1.4$
$x = 3$, $y = 0.8$

While any of these above combination of values will satisfy the given Eq. 8.1, when new equations (i.e., cross validation data) comes in, the new equations may not be satisfied by all these combination of values. By increasing the training data size (equations), you are weeding out too many possible solutions.

If you have more than n equations, you may not get any solution which will satisfy all the equations simultaneously. Consider adding another equation:

$$5x + 10y = 24 \tag{8.3}$$

You will now see that there is no possible value combination for x and y, which will satisfy all three equations. In such a case, whatever value of x and y you choose, there will always be some error. The aim here for you is to choose x and y such that the total error is minimal. With appropriate choice of x and y, made by considering many equations (training data), the error for any new equation (cross validation data) will also be in the same ballpark.

On the other hand, say, you introduce a third variable, z, into all these equations. Assign whatever random coefficient to this third variable for each of these three equations. You will find that it will be possible to find an exact solution to these three equations. Effectively, this third variable resulted in an overfit. It was used to artificially give an exact solution, while the variable may have no real role and, hence, reduced the error on training data but can contribute to an error when a new data (cross validation data) comes in. It is better to not have this third variable and get some error on the original set of three equations.

8.8 Utilizing the Bias vs Variance Information

The first thing you want to do is to measure the error on training data and cross validation data.

If training error is too low, and the cross validation error is high, this indicates a high variance (overfit). If both training error and cross validation errors are high, this indicates a high bias (underfit). Once you know whether you have a case of bias or variance, your choices are:

- High bias
  - Increase the number of features.
  - Add higher-order polynomials.
  - Decrease λ, so that more features survive.
- High variance
  - Increase data, i.e., get more data.
  - Reduce the number of features.
  - Increase λ, so that some features don't survive.

8.9 Derived Data

You have seen that as you increase the order of the polynomial for the variables being considered, you will start getting derived data, such as x_i^2 or x_i^3 or $x_i * x_j$. Sometimes, you might need to consider some other derived values, such as log, exponential, or trigonometric. The question is how do you figure out that you need such derived values. If you consider that the variable being predicted follows a normal distribution, one way of thinking is that all the constituent variables should also follow a normal distribution. So, if your raw data does not follow a normal distribution, you might want to apply some mathematical transformations to see if the data after transformation follows a normal distribution. If after the transformation, the data better fits a normal distribution, this mathematical transformation might give a better result.

8.10 Approach

Start with a very simple model that you can implement quickly. Plot the errors on training data and cross validation data to determine underfit/overfit and decide on whether you want to increase features or increase data.

For specific samples in cross validation data that seem to provide higher contribution to error, look manually at those data much more closely. This manual inspection should yield information as to what is special for these data that causes them to give higher error. This might give indication as to additional features that should be considered.

8.11 Test Data

In Sect. 8.1, you had also created a set of data, called test set (or test data). The entire discussion in this chapter used only training data and cross validation set. Once you have settled down on your algorithm and the features, plot the error on the test data also. For the algorithm trained on training data, the test data is totally new; hence, it should show similar error curve as cross validation data. So plot the error with test data also to confirm that this new and so far unseen (by the training algorithm) data also shows an error characteristic similar to the cross validation set.

The test data brings in value over the cross validation set because the error on cross validation set has been used to fine-tune the algorithm (e.g., determining λ, identifying underfit, overfit, etc.), while the test data is totally new and would be a fresh evaluation of the final algorithm.

Chapter 9
(Artificial) Neural Networks

This chapter will cover the basics of artificial neural networks (ANNs) which are also called multilayer perceptrons. Neural networks are networks of interconnected artificial neurons. Their structure is heavily inspired by the brain's neuron network. A neural network is generally used to create supervised machine learning models for classification, similar to a Logistic Regression model, and is useful in cases where Logistic Regression may not provide reasonable accuracy. Neural networks form the basis of many of the complex applications and algorithms of machine learning. You will subsequently see some of these applications in Chaps. 16 and 17 (Reinforcement Learning), 11 (Recurrent Neural Network Used in Language Processing), and 15 (Convolutional Neural Network). A good understanding of a neural network is necessary to understand these and other applications that have raised so much interest in machine learning. Neural networks are also used in unsupervised learning for compressed representation and/or dimensionality reduction.

9.1 Logistic Regression Extended to Form Neural Network

In Chap. 3, you had learnt about Logistic Regression. Suppose Logistic Regression isn't working for you, and you are thinking of combining two Logistic Regression models in some way to make a more powerful model.

Consider two Logistic Regression models $m1$ and $m2$, connected in series to form a new model. You have the training examples with classification labels. The first model $m1$ will be fed with the input of the training example, and its output is an intermediate value, which is then fed into the next model $m2$. $m2$'s output should match the classification label in the training example. Figure 9.1 shows these two models connected in series.

Since you already know the inputs for $m1$ and the outputs for $m2$, the remaining important question now is how to obtain the intermediate values (i.e., output from $m1$ that is fed as input to $m2$) corresponding to training data. Both these models $m1$

G. Rebala et al., *An Introduction to Machine Learning*,
https://doi.org/10.1007/978-3-030-15729-6_9

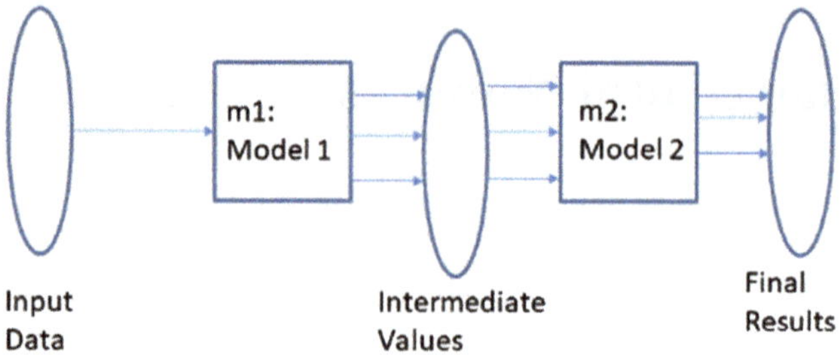

Fig. 9.1 Series connection of two networks performing Logistic Regression

and $m2$ could be trained independently, if you have the intermediate values. The trick here is to iterate on the parameters for both $m1$ and $m2$, in a systematic manner.

The key idea is to tweak the intermediate data going as input to $m2$ when minimizing $m2$'s loss function. That is, from $m2$'s perspective, intermediate data is essentially treated as parameters that can be modified, and not as fixed inputs. So when using gradient descent method, you would have to compute the gradients of intermediate data as well.

Below pseudo code shows how to train the combined model by iteratively training $m1$ and $m2$ *sequentially*, until the loss function on $m2$ becomes sufficiently small on the training data:

```
Initialize m1's parameters to random values
// Tr is the labelled training dataset. Tr.inputs are the training inputs,
 // Tr.outputs are the training outputs
 Run model m1 on inputs Tr.inputs, and store its outputs in Int_val_orig
 Initialize m2's parameters to random values
 While (m2's error function is not too small)
 {
   Train m2 using inputs Int_val_orig and outputs Tr.outputs, while simultaneously adjusting m2's
parameters and Int_val_orig values.
        Store updated Int_val_orig values as Int_val_mod
        Train m1 with inputs Tr.inputs, and outputs Int_val_mod
        run m1 using Tr.inputs, store its outputs in Int_val_orig.
 }
```

If you observe this algorithm carefully, you may discover a way to simultaneously train both $m1$ and $m2$ for faster convergence. If you consider the loss function for the combined model, it is a function of both $m1$'s and $m2$'s parameters. So you can compute the gradient of the combined model loss function for each of $m1$'s and $m2$'s parameters and use gradient descent to simultaneously update both sets of parameters to minimize the loss function.

What you just learnt is essentially an example of an artificial neural network (ANN) with a single hidden layer! You can extend this model by incorporating more intermediate values. For example, you could have three models connected in series so that there are two sets of intermediate values – the output of the first model and the output of the second model. This would translate to an ANN with two hidden layers.

Though this section introduced neural networks as an extension of Logistic Regression, the original idea for a neural network was inspired by studying the brain function.

9.2 Neural Network as Oversimplified Brain

Neural networks were created with the thought that they are mimicking the human brain. Today, with more understanding of the brain, it is known that the neural network is nowhere near close to how the brain works. However, some of the key ideas behind the neural network were derived from a basic understanding of the brain's working – shown in Fig. 9.2.

- Neurons are connected to each other via synapses. The synapses are like electrical wires that transmit electric pulses between neurons.
- A neuron's functionality can be modeled as a multiple-input, single-output function (called the activation function).
- The inputs to a neuron could come from other neurons (through synapses) or primary inputs through sense organs (skin, ear, eyes, tongue, nose). When coming from other neurons, the input reaching here is a function of the output sent by the previous neuron as well as the strength of the synapse carrying that specific signal.
- The strengths of the synapses vary with time. Moreover, one of the ways the brain learns new tasks or new experiences is by increasing or decreasing the strengths of various synapses.

The above basic understanding allowed the formation of a neural network based on the following principles:

- Create an interconnected set of artificial neurons and assign some real valued weights to the interconnections, to represent synapse strengths.
- The network has a fixed set of external inputs feeding into designated neurons within the network and a fixed set of external outputs coming from other neurons in the network. The external inputs are real numbers, and the neural network transforms them into outputs that are also real numbers.
- An ANN could be trained to approximate any function by simply altering the weights of the neuron interconnections. The weight alterations are based on

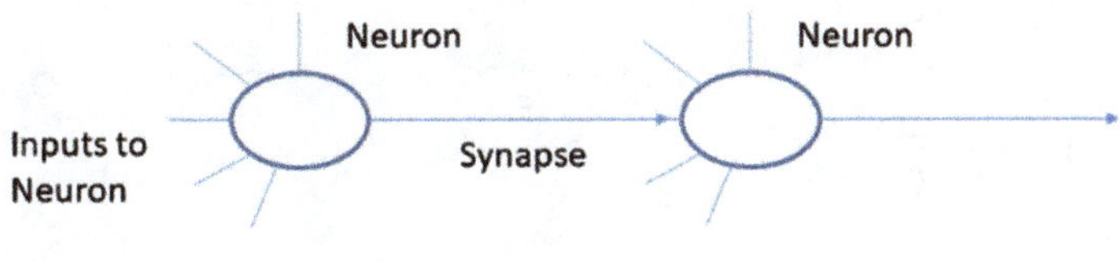

Fig. 9.2 Simplified view of the brain's neural network

several examples of function's input/output behavior, similar to the Linear Regression or the Logistic Regression.
- The neurons perform nonlinear operations.

The rest of the sections in this chapter will go over the details of artificial neural networks including their visual representation, the model, and their training.

9.3 Visualizing Neural Network Equations

Because of the complexity of mathematical formulations in a neural network, it is much easier to represent and understand a neural network in graphical form. And, like all graphs, it is represented as a combination of nodes and edges. For example, the graph in Fig. 9.3 represents an expression $y = m_1{}^*x_1 + m_2{}^*x_2$.

The graph representation of a neural network can be best understood from an example, shown in Fig. 9.4.

- The inputs feed into the first layer of the network. In the example figure, I_1, I_2, and I_3 represent inputs to the neural network and feed into nodes A, B, and C, respectively. The nodes in the first layers do not do any processing. They simply transfer the values to the output side of the node.
- The edges represent the weights. The values at the output of a node get multiplied by the weights, when they reach the input of the next layer. In the example figure, w_1, w_2, etc. represent the weights. So, the value of node A will be multiplied by w_1 when it reaches the input of node D.

Fig. 9.3 Graph representation for $y = m_1{}^*x_1 + m_2{}^*x_2$

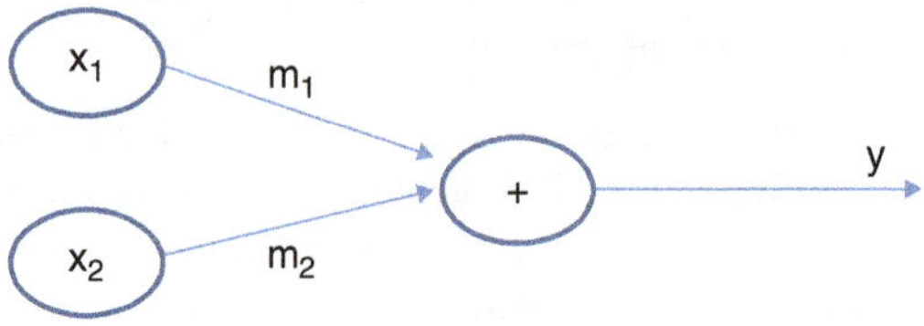

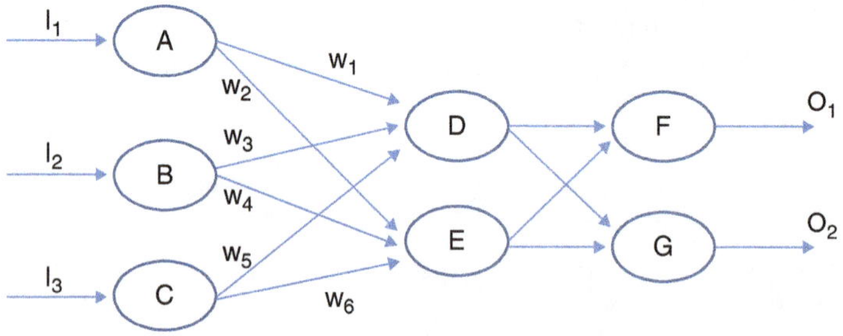

Fig. 9.4 Graph representation of a sample neural network

- Conventionally, the values move from left to right, and hence, the arrows on the edges are often omitted while drawing neural networks, though we have retained the arrows in Fig. 9.4, for the sake of ease in understanding.
- Other than input nodes, all other nodes perform two functions:
  - The first function sums all the values reaching its input. This sum is called *net input* (network input).
  - The second function is the *activation function* applied to *net input. Activation function* is typically a *Sigmoid* function that you had learnt as part of Logistic Regression. However, you may also have other nonlinear functions as activation functions. The output of *activation function* is the output of the node.

  In the given example, at node D, first, the sum is computed: $A{*}w_1 + B{*}w_3 + C{*}w_5$, where A, B, and C represent the values at the output of respective nodes. And then, that sum is given as input to *Sigmoid* function, whose output becomes the node output.

The output of the nodes (i.e., output of the activation functions) is simply called *activation*. Usually, in a classical graph representation, one node represents one function (say a + operator). However, in the world of neural network, the non-input nodes include two functions chained together: summation of incoming values and then an *activation* on the sum.

So, now you should be able to comprehend a neural network graph. Neurons (or nodes) are also called *units* in some literature.

9.4 Matrix Formulation of Neural Network

While graphs are good to visualize, for the training and prediction, you will need to represent the neural network in mathematical form. Matrices provide a relatively compact way of representing and computing with it.

Let $x_1, x_2, \ldots, x_n$ be the n inputs to a node. These inputs can be represented as a column vector: $X = [x_1\ x_2 \ldots x_n]^T$. Similarly, the corresponding weights $w_1, w_2, \ldots w_n$ can be represented as another column vector: $W = [w_1\ w_2 \ldots w_n]^T$. Now, the input to the activation function can be written as X^TW, and the *activation* (i.e., output of the node) y can be obtained by applying the activation function F as $y = F(X^TW)$.

The above explanation was for a single node. However, it is possible to represent the neural network in a matrix form by following a naming convention. Referring again to Fig. 9.4:

- The primary inputs I_1, I_2, and I_3 may be represented as a feature vector.
- The nodes can be named N_{ij}, where i represents the node layer and j represents the node index (or enumeration) within the layer. For example, node A would be called N_{11} (first layer, first node), node B (first layer, second node) will be called N_{12}, node D (second layer, first node) will be called N_{21}, and so on.

- The weights may have names w_{ijk}, where i represents the node layer which will be weighted by this value, j represents the node index within the layer for the originating node, and k represents the node index within the layer for the target node. For example, w_1 (weight starting from layer 1, originating at the first node of the layer and reaching to the first node of the next layer) will be represented as w_{111}; or w_2 (weight starting from layer 1, originating at the first node of the layer and reaching to the second node of the next layer) will be represented as w_{112}. These can be represented as 3D *tensors* (i.e., 3D arrays).
- The 0th input to each layer may often be omitted while drawing. Its value is always 1 and is also called as *bias unit.*

9.5 Neural Network Representation

Figure 9.5 shows another neural network – this time slightly larger. The node and edge labels are often omitted in drawings. Dotted rectangles have been added for explanation only. They are not used during actual neural network representation. One dotted rectangle represents a layer. In practice, all nodes in a layer are arranged vertically in the same line – rather than using rectangular boxes.

Observe that nodes within a dotted rectangle (layer) connect to nodes in adjacent dotted rectangles but not to each other.

The leftmost dotted rectangle is layer 1. This layer has just the inputs to the neural network and no activation function. It is the *input* layer.

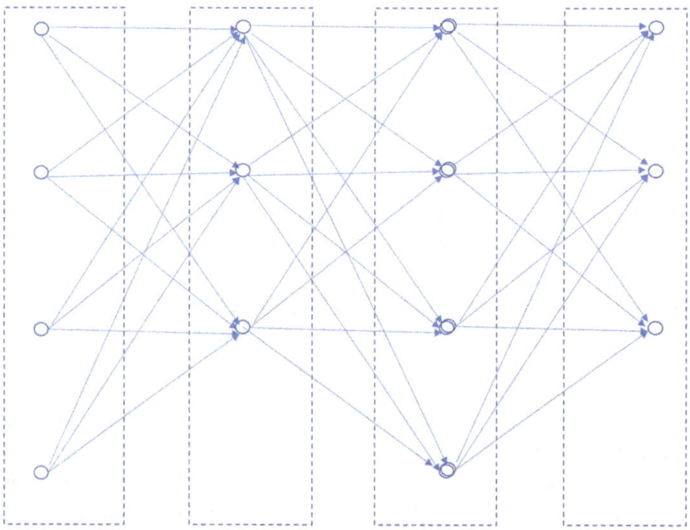

Fig. 9.5 A larger neural network

In all other layers, the nodes represent artificial neurons, and they have an activation function in them.

The rightmost dotted rectangle is the last layer. This layer's activations are the outputs of the neural network. Therefore, this layer is commonly called the *output* layer.

The layers in the middle which are neither the input or output layers are called the *hidden* layers. The network shown in Fig. 9.5 has two hidden layers.

The number of neurons in a layer can vary with layer. For example, in Fig. 9.5, the *input layer* has *four* nodes, the first *hidden layer* has *three* nodes, the second *hidden layer* has *four* nodes, and the *output layer* has three nodes.

Notice that for the network in Fig. 9.5, each layer is fully connected with adjacent layer, i.e., each node in a layer has edges to all nodes in the adjacent layers. Such layers are called *fully connected layers*. It is not necessary for all layers to be fully connected to each other. For example, the convolutional neural networks (covered in Chap. 15) have many layers, and the first few layers (involved in convolution) are not fully connected.

The neural network described so far is known as *feedforward* network, because if you look at its graphical representation, the computation is done layer by layer in one direction – the forward direction from left to right. That is, computation proceeds from the *input layer* to the first *hidden layer* and then from the first hidden layer onto the second hidden layer, and so on until the *output layer*.

The *output layer* of a neural network for classification tasks is usually replaced by *SoftMax* function (described in Chap. 4). That is, for the *output layer*, the activation function would be the *SoftMax*, rather than a *Sigmoid* function.

You now know how to interpret a given neural network. The next step is to learn to design one.

9.6 Starting to Design a Neural Network

Neural networks have many possible topologies. Your neural network design depends on the following factors:

- Number of layers.
- Number of nodes in each layer, which could be a different value for each layer.
- Level of connectivity between nodes of each adjacent pair of layers (fully connected or not)
- The activation functions used in each node (except for the nodes in *input layer* which have no activation function). Choices include but are not limited to *ReLu*, *Sigmoid*, *TanH,* etc. Or, using *SoftMax* for output layer.
- Whether some of the weights are shared between multiple connections (e.g., CNNs share a lot of weights – as you will see in Chap. 15).

- Any feedback edges within the network. *Recurrent Neural Networks* (*RNNs*) (covered in Chap. 11) make use of such feedback. For the purpose of this chapter, assume no feedback.
- Loss function, also known as cost function. Typically, the loss function is always the negative log likelihood function.

Several of the above characteristics define the size, complexity, and overall architecture of the network. Such parameters are called *hyperparameters*. Examples include the number of layers, the number of nodes in each layer, etc. One of the primary tasks in designing a neural network is to explore the space of *hyperparameters* (i.e., try out different combinations of *hyperparameters*) and decide which set of *hyperparameters* is well suited for a model based on training data, available compute capabilities, and other factors. The number of nodes in the *input layer* is simple. It has to be the same as the number of input features, plus one more for the bias unit. The number of nodes in the *output layer* for a classification problem should be the same as the number of classes or categories, so that the outputs are *one-hot* encoded. That is, when a neural network predicts or *infers* a category, the corresponding output will be 1, but all other outputs will be zero.

9.7 Training the Network

Neural network training is similar to training any parameterized model such as Logistic Regression, though neural network has many more *hyperparameters*. The pseudo code for basic training approach is given as:

```
Choose a set of hyperparameters (such as number of layers, the activation functions etc)
begin loop // Iterate 10s to 100s of times depending on problem and time constraints
    Train the neural network model for correct weight (details later)
    If the loss function is low – you have a trained network and exit the loop
    If the loss function is high – modify your set of hyperparameters
end loop
```

The concept of finding the weights for a neural network is very *similar to Gradient Descent* algorithm discussed in Chap. 3. Unfortunately, closed-form solutions (such as *Normal Equation* method for Linear Regression) do not exist for finding the weights for a neural network, and iterative methods based on Gradient Descent are the only available option. The training is typically done with a large number of labelled training examples.

Gradient Descent involves finding partial derivative (i.e., gradient) of the overall objective function (loss function in our case) with respect to each of the weights. This gradient value determines how you adjust the corresponding weights. The *chain rule* of differential calculus will be useful in computing the gradient values for each of the weights.

9.7.1 Chain Rule

The chain rule essentially allows computing the derivative of a function as a product of two or more, possibly simpler derivatives. Let Z be a function (say F) of Y; and let Y be another function (say G) of X, i.e., $Z = F(Y)$ and $Y = G(X)$.

Now, if you want to find the derivative of Z with respect to X, the easiest method is as given in Eq. 9.1. This *chain rule* works only if functions F and G are individually *differentiable* with respect to Y and X, respectively:

$$\frac{dZ}{dX} = \frac{dZ}{dY} * \frac{dY}{dX} \tag{9.1}$$

where

$\frac{dZ}{dY}$ represents $F'(Y)$, i.e., derivative of Z with respect to Y
$\frac{dY}{dX}$ represents $G'(X)$, i.e., derivative of Y with respect to X

A crude way to remember this rule is to imagine that *dy* cancels out since it occurs in the numerator and denominator of the product – though this is not the real mathematical explanation! Here, you are depending on an intermediate value Y.

Consider a simple example to illustrate the chain rule. Consider the two functions shown in Eqs. 9.2a and 9.2b:

$$Z = Y^2 \tag{9.2a}$$

$$Y = 2X + 3 \tag{9.2b}$$

Using the chain rule, $\frac{dZ}{dX}$ can be computed as $\frac{dZ}{dY} * \frac{dY}{dX} = 2Y * 2 = 4Y = 4(2X + 3) = 8X + 12$. For this simple example, verify this result by rewriting Z as a function of X and directly computing the derivative of Z w.r.t X.

You can create such cascades of products using any number of intermediate variables and functions. Consider related variables and functions such as:

- $y_4 = F_4(y_3)$
- $y_3 = F_3(y_2)$
- $y_2 = F_2(y_1)$
- $y_1 = F_1(x)$

You can find $\frac{dy4}{dx}$ by applying the formula given in Eq. 9.3, as derived from the cascaded application of the chain rule:

$$\frac{dy4}{dx} = \frac{dy4}{dy3} * \frac{dy3}{dy2} * \frac{dy2}{dy1} * \frac{dy1}{dx} \tag{9.3}$$

For our purposes, the functions involved have multiple-input variables. Hence, the above concept has to be extended to multiple variables. Let Z be a function F of three variables Y_1, Y_2, and Y_3, i.e., $Z = F(Y_1, Y_2, Y_3)$. Each of these three variables in turn is a function of several other variables, including X, as given below. The ... indicate the presence of other variables:

- $Y_1 = G_1(X, \ldots)$
- $Y_2 = G_2(X, \ldots)$
- $Y_3 = G_3(X, \ldots)$

The need is to find the partial derivative of Z w.r.t X, i.e., $\frac{\partial Z}{\partial X}$. The partial derivatives of Z w.r.t Y_1, Y_2, and Y_3 are given by Eqs. 9.4a, 9.4b, and 9.4c:

$$\frac{\partial Z}{\partial Y1} = \frac{\partial F(Y_1, Y_2, Y_3)}{\partial Y_1} \tag{9.4a}$$

$$\frac{\partial Z}{\partial Y_2} = \frac{\partial F(Y_1, Y_2, Y_3)}{\partial Y_2} \tag{9.4b}$$

$$\frac{\partial Z}{\partial Y_3} = \frac{\partial F(Y_1, Y_2, Y_3)}{\partial Y_3} \tag{9.4c}$$

Similarly, find the partial derivatives for each Y_1, Y_2, and Y_3, with respect to X. *Now, to find the partial derivative of Z with respect to X, you will need to consider the contribution made by each of the intermediate variables*, Y_1, Y_2, and Y_3, and then, those individual contributions will need to be added up. The overall expression for such a partial derivative is given by Eq. 9.5:

$$\frac{\partial Z}{\partial X} = \frac{\partial F}{\partial Y_1} * \frac{\partial Y1}{\partial X} + \frac{\partial F}{\partial Y_2} * \frac{\partial Y_2}{\partial X} + \frac{\partial F}{\partial Y_3} * \frac{\partial Y_3}{\partial X} \tag{9.5}$$

where

∂F is used to represent $\partial F(Y_1, Y_2, Y_3)$ – for the sake of brevity

9.7.2 Components of Gradient Computation

You need to compute the gradient/slope of the loss function w.r.t each of the weights for a given labelled training data. The value of the corresponding weights would then be adjusted accordingly. Let the loss function be represented by J. Refer to Fig. 9.6, which shows a specific neuron on the output layer. Y_i denotes the output value (i.e., activation), and S_i denotes the net input (i.e., weighted sum of all incoming net values).

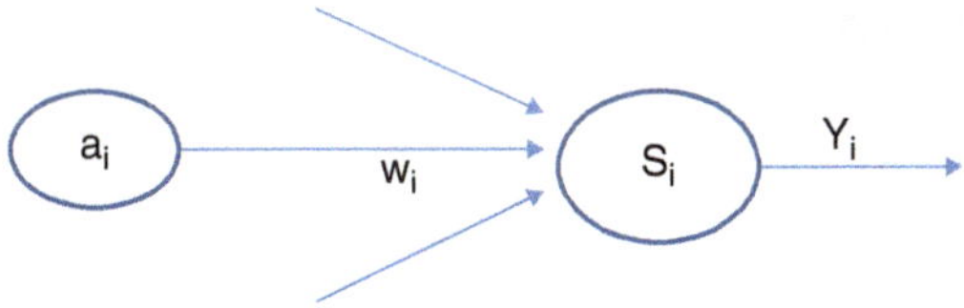

Fig. 9.6 A specific neuron on the output layer

Table 9.1 Derivative values for activation functions

Activation function	Derivative
Sigmoid	Activation value * (1 – activation value)
ReLu	1, if net input value >0 0, if net input value <0 Undefined, if net input value is 0 (you can take 0)

First, compute the gradient of J with respect to Y_i, i.e., $\frac{\partial J}{\partial Y_i}$. Using the chain rule, you can compute $\frac{\partial J}{\partial S_i}$ as the product $\frac{\partial J}{\partial Y_i} * \frac{\partial Y_i}{\partial S_i}$. Recall that the activation function is a function of a single variable. Hence, $\frac{\partial Y_i}{\partial S_i}$ will be a simple derivative of the activation function, rather than a partial derivative.

It is evident from the above procedure that for a given output neuron, you can compute the gradient of the loss function w.r.t its net input as the product of two terms:

- Gradient of the loss function w.r.t activation
- Derivative of the activation function

The derivative of an activation function will depend on the actual function chosen for activation. Table 9.1 provides the derivatives for two popular activation functions.

Net input S_i is expressed as $\sum w_i * a_i$, where a_i denotes the activation of a node in the previous layer and w_i denotes the corresponding weight. $\frac{\partial S_i}{\partial w_i}$ evaluates to a_i, since all other terms in the summation are independent of w_i.

Applying the chain rule once again, $\frac{\partial J}{\partial w_i}$ can be given as the product $\frac{\partial J}{\partial S_i} * a_i$. So, you can compute the gradient of J with respect to any weight, given the gradient with respect to the corresponding net input. Referring again to Fig. 9.6, given the gradient of J with respect to S_i, you can compute the gradient with respect to w_i, a weight going into the output layer. Now you need to ensure that you have the gradient with respect to the activations of hidden layer neurons also – so that you can find the gradient with respect to the corresponding weights.

Consider the last hidden layer (i.e., the layer connected to the output layer), as shown in Fig. 9.7. S_1 and S_2 are the net inputs to the neurons of the output layer. Recall that you have already calculated the gradient of J with respect to these S_i (for the output layer).

Considering the node denoted by N_i in Fig. 9.7, its contribution to S_1 and S_2 is given by Eqs. 9.6a and 9.6b:

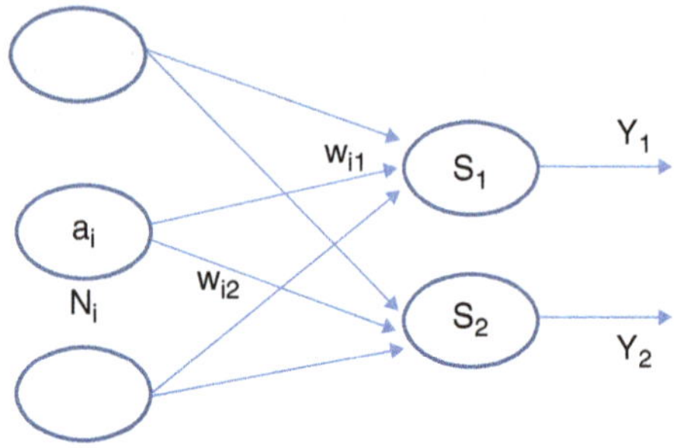

Fig. 9.7 The last hidden layer

$$S_1 = \ldots + a_i * w_{i1} + \ldots \tag{9.6a}$$

$$S_2 = \ldots + a_i * w_{i2} + \ldots \tag{9.6b}$$

The loss function is a function of the neural network outputs: Y_1 and Y_2, which in turn are directly derived from S_1 and S_2 by applying the activation: Y_i = activation (S_i).

With the understanding so far, the gradient of J with respect to a_i is given by Eq. 9.7a, further simplified to 9.7b:

$$\frac{\partial J}{\partial a_i} = \frac{\partial J}{\partial S_1} * \frac{\partial S_1}{\partial a_i} + \frac{\partial J}{\partial S_2} * \frac{\partial S_2}{\partial a_i} \tag{9.7a}$$

$$= \frac{\partial J}{\partial S_1} * w_{i1} + \frac{\partial J}{\partial S_2} * w_{i2} \tag{9.7b}$$

Now that you know how to compute derivatives for activations on the last hidden layer (say layer $N - 1$) and the weights from that layer to the output layer, you can use similar methods to compute derivatives of activations of layer $N - 2$ and weights feeding layer $N - 1$. You can extend this method to compute derivatives of all weights in the neural network.

Sometimes the weights can be shared by multiple edges, usually in the same layer, such as in CNNs. In other words two or more distinct edges may be represented by the same weight variable. In such cases, the gradients can be computed for each of those edges as usual and then added up to get a single gradient for that weight.

9.7.3 *Gradient Computation Through Backpropagation*

Now that you understand each of the individual pieces, let us put it all together, as shown in the following pseudo code:

```
Initialize L = index of output layer // L = 1 for input layer
Initialize each of the weights to random values

For a given set of input features, compute all the outputs and hence determine J.
While (L ≥ 2) {
For each neuron N in the layer L {
Compute Gradient of J wrt N's net input (i.e. ∂J/∂S)
Compute Gradient of J wrt each weight feeding into N's input (i.e. ∂J/∂W)
}
For each neuron N in the layer L - 1 {
Compute Gradient of J wrt N's activation (i.e. ∂J/∂a)
}
Decrement L
}
```

The algorithm computes the gradient of J with respect to activations, net inputs, and weights. For each such gradient computation, it reuses the gradient of the previous stage and multiplies the derivative of just this stage. Such reuse has been possible due to the chain rule. Such computation of gradient starts from the output activations and keeps propagating backward through hidden layers – till it reaches the weights originating from the input layer. Since the gradient computation propagates backward, this process is called *backpropagation*.

Since this concept of backpropagation is slightly complex to conceive, look at it once again, conceptually, using the network shown in Fig. 9.5 as an example. Starting with the random assignment of weights, you predict the values for the outputs based on a given input data. Considering the error at the output and the activation function, you update the values reaching the inputs of the nodes on this output layer. These values can be updated by updating the weights connecting the nodes on layer 3 to this output layer or by updating the values coming out of the nodes in layer 3. So, update both: the weights and the output of the nodes on layer 3. Now, come to layer 3. You now know the values that should be on the output of nodes on this layer. So, you can compute the input on the nodes on this layer. And once again, you update the weights feeding into this layer, as well as the nodes on layer 2. You keep updating these values backward. That will complete one iteration.

Now, use the next labelled data to predict the values from the output (forward propagation), compute error, and further refine the weights through backpropagation.

9.7.4 Updating Weights

Once the gradients are computed for each weight, adjust these weights – based on the sign of their slope, the magnitude of the slope, and the learning rate, using the same concept of *New weight = Old weight – Learning Rate * Gradient*.

Satisfy yourself that when the weights are updated, the various activation values get updated automatically (with the activation functions remaining unchanged).

Thus, as part of training, only the weights have to be updated. The gradient (derivative) with respect to activation values needs to be determined during backpropagation solely to exploit the chain rule.

Instead of a simple learning rate, you can use more advanced methods described in Sect. 18.3.

9.8 Vectorization

The explanation has been provided as if you will compute for one neuron at a time or one weight at a time. In reality, you will exploit vector algebra and matrix operations, as explained in Sect. 9.4.

9.9 Controlling Computations

You have seen that for computing just one round of updates to the weights (based on Gradient Descent), you have to do a lot of computations. For a large network (used in deep learning), you might have hundreds of layers and tens of thousands of nodes – resulting in millions of weight parameters (i.e., weights). For a single iteration, you need to compute the slope for all the weights.

Usually, training datasets can have a large number of examples (e.g., tens of millions for image classification). Therefore, it will be too slow to train the neural network one example at a time. The most popular method of training neural network is using Stochastic Mini Batch Gradient Descent, with mini batch size being a hyperparameter. This provides a good balance between compute effort (each mini batch has a small number of examples, limiting the computation required for forward propagation and backpropagation) and convergence (faster convergence requires larger mini batch sizes).

9.10 Next Steps

You are now ready to embark on the journey into deep learning.

For now, feel free to create neural networks using library functions in any of the popular machine learning packages, and play with various hyperparameters to see how the model accuracy and training time vary with network topology. You can use one of the stock-related data and problems mentioned in earlier chapters.

Once you have gotten a feel of how neural network behaves, you can try to implement neural networks and backpropagation from scratch, to gain higher level of confidence and expertise.

Chapter 10
Natural Language Processing

Natural language processing (NLP) is understanding and interpreting human languages, spoken or written, using machine processing. NLP is useful in a variety of applications including speech recognition, language translations, summarization, question responses, speech generation, and search applications. NLP is an area of research which has proven to be difficult to master. Deep learning techniques have started to solve some of the issues involved in natural language processing.

Human speech has constructs that assume prior knowledge of the world or common sense in understanding and interpreting context of the text. The context, under which the words are used, can derive meaning and makes it difficult for machine processing to apply common sense to the sentences. For example, "Candy is sweet," based on the context, "Candy" could mean as a name of a person or a piece of confection.

10.1 Complexity of NLP

Natural language processing is based on linguistics and statistical inference techniques. Most of the original techniques for NLP were based on a complex set of handwritten rules that correspond to grammar, compound word formation, ontologies, and extracting context.

Since the early 1980s, the focus on NLP has shifted to statistical models based on the available text in the real world. The increase in computational power provided a significant boost to complex statistical and probabilistic models, based on the available samples of real-world text or *corpora*.

Some of the properties of language that make it difficult to process are lexical ambiguity, acronyms, syntactic ambiguity, noun compound interpretations, and differences in spoken and written structures. Consider some of the problems posed by such language properties.

G. Rebala et al., *An Introduction to Machine Learning*,
https://doi.org/10.1007/978-3-030-15729-6_10

- Acronyms: It is very common to use acronyms in any field of study even if we do not consider the text messaging-based English. NLP could mean neurolinguistic programming or natural language processing in the realm of language understanding and analysis. In the context of programming, linear programming counterpart as non-linear programming could also be NLP in that context. Some of the acronyms can have dozens of expansions, and more are invented all the time.
- Lexical ambiguity: Words with same spelling can have different meanings depending on the context of usage. Most languages have this property making it difficult to assign a meaning without the context in which they are being used. Consider the statements:

    1. *Wait at the door, I will park the car.*
    2. *Wait at the door, I will fetch the car from the park.*

From the first statement, you would understand that the *park* meant to place the car into parking spot. The second statement is understood as the car is near a *park*. The context of the word usage or the topic of the word category will give it the meaning under the conditions.

- Syntactic ambiguity: A sentence spoken is usually not very precise as common sense is assumed when spoken and even in written text. The punctuation can also make a significant difference in the meaning of what is being expressed. Consider the example of a well-known statement by Grouch Marx – "One morning, I shot an elephant in my pajamas. How he got into my pajamas, I don't know."

A commonsense understanding of the statement "I shot an elephant in my pajamas" would be to assume that I am in my pajamas; however, the same statement could also mean that the elephant is wearing my pajamas while I shot it. Without the commonsense understanding of the world around us, both of the meanings of the statement are equally likely, however unlikely they both may seem with common sense.

- Compound words: Two or more words make up another compound word that is commonly used, for example, car park, soap opera, dry clean, and cleaning products. The compound words are more common in Sanskrit-derived languages based on their grammar and in Germanic languages. Occurrences of the compound words in the body of text are sparse, and the context under which the words are compounded is hard to learn from the text. The meaning of a compound word may not be directly related to the individual words making the compound word. One such example is *hot dog*.

The next section discusses some of the common techniques used in natural language processing to address some of the complexities in natural languages and understanding the context.

10.2 Algorithms

Attempts to process natural language are made using the rule specification of grammar, statistical, and probabilistic models and, in the recent past, neural networks. NLP had not seen a great success in the twentieth century. Significant success is seen using neural network models in the past 10 years. This section discusses some of the pertinent algorithms.

10.2.1 Rule-Based Processing

Rule-based systems are useful when there is small amount of data for learning statistical-based models or to implement a targeted small system that has well-known structure. Rules are also useful in providing user feedback to statistical models to improve the results.

In natural language processing, rule-based systems were some of the first models developed for checking grammar, spelling correction, and resolving compound words. The systems were built as knowledge-based systems with extensible rule specification capabilities.

Rule-based systems generally have knowledge trees with system matching possible rules with the words and sentences encountered. The rules are constructed in the form of a tree, with traversal in the tree as the words are encountered in a sentence. If the path does not match any rules for the next word, then the rules are retracted to trace another path in the tree.

This approach results in success with smaller set of data. However, when the number of rules required is quite large, then the construction and maintenance of the rules become untenable. Most of the modern systems use statistical and machine learning techniques with rule specification to enhance the results. For example, you could specify rules for determining white spaces. Such rules (for white space determination) could be different based on applications as some applications may require to preserve parenthesis and others do not. Another example could be the search results to display score for a baseball game when searched for a baseball team.

10.2.2 Tokenizer

A body of text is a sequence of characters that is converted into a string of characters represented as a token, based on a method to demarcate strings. The process of constructing tokens from the text is called tokenization. The resulting tokens are then sent for further processing for storage and retrieval.

A simple tokenization of English text is based on white space as a delimiter. A white space is a character or series of characters that represent a space, tab, line ending, page breaks or consecutive elements of whitespace set.

Consider the statement – *A quick brown fox jumps over the lazy dog*. Constructing tokens on this statement gives us tokens of *a*, *quick*, *brown*, *fox*, *jumps*, *over*, *the*, *lazy*, and *dog*.

Tokens are usually stored in the format that preserves the information present in the statement, like the order of the words to prevent loss of information in storage for processing. Tokenizers are generally not responsible for syntax checking or extracting semantics from the generated tokens.

A common and popular model of tokenizer in natural language processing is called *N-gram* tokenizers. *N-gram* method uses *n* consecutive characters to form a token. The tokens formed by this method may not be present in the dictionary. They also have an advantage of including unfamiliar words and context of consecutive words in a document.

In constructing *N-grams*, most processors remove white spaces as they do not add much information. However, white spaces can also be considered while constructing *n-grams*. As an example, forming 4-*gram* tokens for phrase "Majestic mountain" will give the following:

Maje, *ajes*, *jest*, *esti*, *stic*, *tic* , *ic m*, *c mo*, *mou*, *moun*, *ount*, *unta*, *ntai*, *tain*

Natural language processing models make an assumption that the probability of occurrence of the next token can be computed based on past *n* tokens alone. This simplifies language model and still is effective in predicting the probabilities of the words in forming the sentence.

10.2.3 Named Entity Recognizers

In NLP, recognizing nouns or entities and placing them in proper categories, like names of people, names of organizations, names of places, currencies, quantities, etc., are critical in extracting the context of the text. When one encounters Tesla, it can be recognized as a name of a person or name of a motor company or a measuring unit for magnetic fields. Recognizing entities correctly is critical for extracting meaning from documents.

A common approach to named entity recognition is to provide a rule-based approach. Machine learning models provide better classification with newer terms. A semi-supervised learning with a machine learning model, augmented with training and feedback, has shown to provide a very high accuracy.

The models are generally domain specific as they are found to be brittle, when applied across domains. The models, trained for medical field or military communications, do well for their respective domains, and applying them across domains reduces their precision.

10.2.4 Term Frequency-Inverse Document Frequency (tf-idf)

Searching documents is to find the best document that matches the terms being used in the search context. The most common technique is to use the importance of the word in a document by computing how many times the term occurred in the document.

Term frequency refers to the frequency of occurrence of a term in the document. An easy surmise would be that the higher the frequency of the occurrence of the term, the higher the relevance of the document. However, in large documents compared to smaller documents, the term may occur more often simply because of the fact that the document is long, and hence the term occurs more frequently. To reduce this bias, the length of the document is considered to reduce the effect of larger documents:

$$Tf = \frac{\text{Number of times term occurred}}{\text{Total number of words}}$$

Inverse document frequency (*Idf*) identifies the frequency of occurrence of a term in all the documents. This analysis helps in identifying terms that are common among multiple documents and hence does not add much distinguishing information by using the term. For example, searching based on "the" or giving addition weight to "the" because it occurs at high frequency in a document is not useful as opposed to term "find". Equation 10.1 gives the formula for computing *Idf*, and Eq. 10.2 shows the formula for computing *tf-idf*:

$$Idf = \log(N/\text{num of documents where } t \text{ is found}) \tag{10.1}$$

$$Tf\text{-}idf = tf(t,d) * idf(t,D) \tag{10.2}$$

where,

t is the term used in search
d is the document under search
D is the collection of all documents
N is the total number documents

A value of *tf-idf* is high if the term frequency is high in a document and the occurrence of the term in the entire document collection is low. However, if the frequency of term in the document is high and also high across all the documents, then the weight will be close to 0. This will make sure that the common terms are given low score and ones that are frequent in one document and sparse in the corpus are given high score.

10.2.5 Word Embedding

Word embedding is way of mapping words into vectors of numbers that preserve a form of syntactic and semantic relationships between words. Machine learning algorithms or statistical processes require number formats to process large amounts of data and are not good at string processing.

A simplest form of word embedding for a document is to construct a vector with word counts. Say, there are n unique words in the document, then construct a vector with n dimensions, where each dimension represents a word and the strength of each dimension by frequency of occurrence of the word.

Consider the two text structures as below for vector construction:

D1: "Park the car near the park."
Unique words in the text are *park*, *the*, *car*, and *near*.

D2: "She is at the park."
Unique words in the text are *she*, *is*, *at*, *the*, and *park*.

Let $V1$ and $V2$ be the vectors representing $D1$ and $D2$, respectively. Table 10.1 shows how these vectors are formed.

The vectors can also be constructed by using *tf-idf* measures instead of simple word counts. This will reduce the impact of common words and increase weights for words that are infrequent in the documents.

Another commonly used method for vector construction is to use covariance matrix representation of the terms in the documents. This method preserves relationship between words as the words that occur together are given higher weights.

A covariance count fixes context by a certain number of words and counts the frequency of occurrence of group of words. Consider two sentences $D1$ and $D2$ as:

D1: *This is a perfect weather for sailing.*
D2: *This weather is perfect for fishing.*

Table 10.2 illustrates counting using covariance matrix with the number of words for context fixed at 2, hence using two words surrounding the current word as relevant context.

From the covariance matrix, you can see that the "perfect" and "weather" have a relatively high frequency of 2. Adding more documents to the corpus with similar occurrence will increase the context of occurrence of adjectives like *perfect* to the *weather*. Removing stop words from the matrix like "is," "a," and "for" can clearly show the relevance of the adjective to *weather* by removing the noise from stop words as shown in Table 10.3.

Table 10.1 Formation of text vectors

	Park	The	Car	Near	She	Is	At
V1	2	2	1	1	0	0	0
V2	1	1	0	0	1	1	1

Table 10.2 Covariance matrix for D1 and D2

	This	Is	A	Perfect	Weather	For	Sailing	Fishing
This	0	2	1	0	0	0	0	0
Is	2	0	1	1	1	1	0	0
A	1	1	0	1	1	0	0	0
Perfect	0	1	1	0	2	2	0	1
Weather	0	1	1	2	0	1	1	0
For	0	1	0	2	1	0	1	1
Sailing	0	0	0	0	1	1	0	0
Fishing	0	0	0	1	0	1	0	0

Table 10.3 Reduced covariance matrix

	Perfect	Weather	Sailing	Fishing
Perfect	0	2	0	1
Weather	2	0	1	0
Sailing	0	1	0	0
Fishing	1	0	0	0

10.2.6 Word2vec

Word2vec is an algorithm developed by Google under Mikolov et al. that outperforms algorithms like latent semantic analysis in finding the semantic relationship between words. *Latent semantic analysis* uses word distributions within the paragraphs of documents to find the semantic relationship between words that occur together. This algorithm is not discussed in this book.

Word2vec is a two-layer neural network model trained to provide weights for associating words and reconstruct context for the words. Word-embedding vectors generated by *word2vec* are positioned close to each other if they share a context in the corpus.

Word2vec utilizes either *continuous bag of words* (CBOW) (explained in Sect. 10.2.6.1) or continuous skip-gram algorithms (explained in Sect. 10.2.6.2) to produce vector representation of words. The order of context words is not important in the *bag-of-words* algorithm, whereas the *skip-gram* algorithm uses the context to predict surrounding words. *Skip gram* does better in predicting infrequent word contexts, whereas the *bag-of-words* algorithm is faster.

10.2.6.1 Continuous Bag of Words

Continuous *bag-of-words* algorithm works by defining output as the bag of words surrounding the input word in the text and input as the vector of the word counts in the bag of words. For example, in the sentence "Children are playing in the park," the input vectors will be as shown in Table 10.4 when selecting one context sample.

Table 10.4 Input to CBOW

	Children	Are	Playing	In	The	Park
Input 1	1	0	0	0	0	0
Input 2	0	1	0	0	0	0
Input 3	0	0	1	0	0	0
Input 4	0	0	0	1	0	0
Input 5	0	0	0	0	1	0
Input 6	0	0	0	0	0	1

Table 10.5 Output from word2vec neural network

	Children	Are	Playing	In	The	Park
Output 1	0	1	0	0	0	0
Output 2	1	0	1	0	0	0
Output 3	0	1	0	1	0	0
Output 4	0	0	1	0	1	0
Output 5	0	0	0	1	0	1
Output 6	0	0	0	0	1	0

Neural network will be trained with the expected output shown in Table 10.5 for each of the inputs.

After training the neural network with the corpus, by inputting a word or collection of words, one will get probabilities of the context words that correspond to the input words. For example, when the input word is “apple,” you could get “mango” and “Samsung” as output words that have high contextual relationship. When multiple words are given as input, the context will be finer to identify the topic and hence give higher probability to “mango” as output, if the context was the fruit, *apple*.

10.2.6.2 Skip-Gram Model

Skip-gram model predicts the expected word given the current context. Unlike the *continuous bag-of-words* algorithm, the order of the words is considered in training the algorithm. The probabilities of context words in the expected order are incorporated in the model.

Skip-gram input is constructed by skipping words in the input text among consecutive words. Using the same text example, “Children are playing in the park,” the skip grams for 1 skip are as follows:

1-skip bi-grams = {*children are, children playing, are playing, are in, playing in, playing the, in the, in park, the park*}

Equation 10.3 gives the number of such skip-gram words formed for an n word sentence with k skips:

$$\text{Number words} = (k+1)(k+2)(3n-2k-6)/6 \tag{10.3}$$

Hence, there can be large number of input data generated from the text. However, the skip grams are generated as part of a sentence and are not allowed to extend between sentences.

Training the neural network similar to the continuous bag of words with the skip-gram set of outputs for each of the input word generates the weights for each of the output words. You can derive probabilities of expected words from the weighs assigned to the output.

For a text of 1000 words, with 100 features, the number of computed weights is 100,000. Larger texts and increased features make this computation prohibitive. There are multiple techniques used in training to reduce the set of weights to compute in the result set. These are not covered in this book.

Chapter 11
Deep Learning

Deep learning has seen significant success in recent time with applications like speech recognition, image processing, language translation, and list goes on. Deep neural networks in general refer to neural networks with many layers and large number of neurons, often layered in a way that is generally not domain specific. Availability of compute power and large amount of data has made these large structures very effective in learning hidden features along with data patterns.

Convolutional neural networks (CNNs) and their success in image recognition problems are discussed in Chap. 15. This chapter discusses *Recurrent Neural Networks* and their derivative *long short-term memory* (LSTM) network. LSTMs are used extensively in natural language processing, language translation, video games (as policy network in reinforcement learning), and many other complex machine learning problems that require decisions to be taken based on past history of inputs.

11.1 Recurrent Neural Networks

Recurrent Neural Networks (RNNs) are models of neural networks that are best suited for processing sequential data. A series or sequence data can be seen in multiple problem domains with time series data or simply a sequence of data that has repetitive properties.

Sequence data can be seen in DNA data that can have a sequence of thousands of base proteins. For example, the DNA in the largest human chromosome consists of 220 million base pairs that have repeating subpatterns. Speech recognition requires modeling time series data with individual speech patterns. *Natural language processing* (NLP) is an area where sequences are critical in identifying the matching word. Consider the following sentence recognized by speech processing, "kids went to play in the bark." Using the sequence of the words in the sentence, the sequence model can then determine "park" has higher probability than "bark" in the context of

G. Rebala et al., *An Introduction to Machine Learning*,
https://doi.org/10.1007/978-3-030-15729-6_11

children and *park* and correct the sentence as "kids went to play in the park." NLP sequence models are used in speech recognition, language translation, sentiment classification, name entity recognition and for generating text in chatbots.

Simple neural networks' construct models assume that the input vectors are independent of each other. Neural networks only consider the current input vector and apply the model on the vector to classify or predict the output. This approach ignores much of the information present in the context for accurate detection or prediction.

To illustrate simple neural network approach for modeling named entity recognition, consider the sentence, "Harry and Sally went to a movie after dinner." In this sentence, the words "Harry" and "Sally" should be identified as named entities. A common approach to process natural language data is to input the sentence as one input for each word, encoded as one-hot vector to a neural network. The dimension of one-hot vectors is the number of unique dictionary words or the number of words in the entire corpus. Each position or component of a one-hot vector represents a binary state of 1 for the presence of a word from the corpus, associated with that position or the binary state of 0 for the absence of a word from the corpus, associated for that position. The current word position is represented as 1, and the rest of the positions are represented as 0. The corpus size of 50,000 words is fairly common and is also considered small as some of the domains with *n-gram* processing can use a corpus with more than a million words. So, the sentence under consideration will produce nine one-hot vectors with each vector the size of the corpus.

The representation for the corpus is shown in Eq. 11.1 with all the words known in sorted order. Equations 11.1a, 11.1b, and 11.1c show the corresponding input and output vectors:

$$\text{Corpus} = [\text{a an} \;\ldots\; \text{after} \;\ldots\; \text{harry} \;\ldots\; \text{hally} \;\ldots\; \text{went} \;\ldots\; \text{zyzzogeton}] \tag{11.1}$$

$$X = [\text{Harry and Sally went to a movie after dinner}] \tag{11.1a}$$

$$X_i = [0\,0 \ldots 1\,0 \ldots 0\,0] \tag{11.1b}$$

$$Y = [1\,0\,1\,0\,0\,0\,0\,0\,0] \tag{11.1c}$$

where

X is the input vector with all the words in the sentence
X_i is the ith one-hot vector of size $1 \times N$ given as input
i ranges from 1 to 9, number of words in the sentence
N is the number of words in the corpus
Y is the output vector classifying 1 for named entity, 0 for all other words

The simple neural network model will take the input with N nodes where N is the size of the dictionary and a binary output of 1 for the named entity and 0 otherwise. Each input vector is a one-hot vector X_i for each of the words, and the output is the binary decision of true or false for the named entity. Typically, due to large size of the corpus running into milliions of words, hot vector sizes are very large. This makes the input layer of the neural network very large causing problems in resource requirements. Word embedding techniques like Word2Vec are used to convert hot vectors into smaller vector sizes which are then used as input to the neural network.

In this approach each of the inputs is independent of each other and does not use information on the position of the words. Sequence data has additional information based on the sequence which requires models that use more than the current input vector. To model information based on the positions, the entire sentence can be input to the neural network instead of word-based one-hot vector.

However, because of the position-based identification of the words during training, the input sentence needs to be sorted and hence will lose position-based information. Since words can have multiple meanings with the same spelling, recognizing the context of word usage is lost. Consider the text in example, "West went to the Eastern delight restaurant with his friends." Here *West* can be recognized as a name of a person which is different than usage in "Sun sets in the West." Using sorted one-hot vectors will not be able to differentiate the usage of the word *West* as it has no contextual connection with rest of the sentence or has multiple contexts with preprocessing with word embeddings.

Recurrent Neural Network (*RNN*) models solve the problem of sequence modeling by providing ability to remember past data and process the current vector based on the inputs that are before or after the current vector in the sequence.

*RNN*s have been very successful in solving context recognition and repetitive pattern recognition problems. For example, in a natural language processing, the recognition of word context requires looking at other words in the sentence. Consider the example, "West went to the Eastern delight restaurant with her friends." In this sentence, the name *West* refers to a person and a female because of the use of pronoun "her," whereas it will refer to a person and male if "her" is replaced by "his," and both are valid uses depending on the context.

Recurrent Neural Network (*RNN*) architecture makes the construction of such neural networks simple and efficient. As the term recurrent in the name implies, the task structure is repeated multiple times in the computation with definition of task done only once. An intuitive understanding is that the past inputs are memorized with weighted values combined with the current input for prediction.

11.1.1 Representation of RNN

A typical RNN architecture consists of multiple neural networks, with sequential inputs given to each of the networks along with processed data from the previous sequence. A simple neural network can be represented as shown in Fig. 11.1.

Fig. 11.1 Simple neural network

From Neural Network chapter, a simple network takes input vector X and output Y with one or more hidden layers. The neural network in Fig. 11.1 shows X is an input vector, W_{ax} is the weight vector from the first layer, W_{ay} is the weight vector from the last layer, and Y' is the output. The network state can be represented as in Eq. 11.2:

$$H_i = g(W_{i-1i}\, H_{i-1} + b_i) \tag{11.2}$$

where

H_{i} is the state of layer i
i is the index of the hidden layer
W_{i-1i} is the weight vector between layers $i - 1$ and i
b_i is the bias term
g is the activation function, usually a Sigmoid

The output of the neural network can be represented by Eq. 11.3:

$$Y' = g\left(W_{iy}\, H_y + b_y\right) \tag{11.3}$$

where

H_y is the state of the last layer
Y' is the output of the neural network
W_{iy} is the weight vector to the last layer from the previous layer
b_y is the bias term of the last layer
g is the activation function, usually a Sigmoid

RNN is built on simple neural network to take advantage of contextual information that is present in the sequence data. Intuitively, every input of the sequence creates a memory state that is computed from current and past inputs. This memory along with the next input vector is used to compute the output of the next sequence. *RNN* computes the state of the network with each input of the sequence and passes

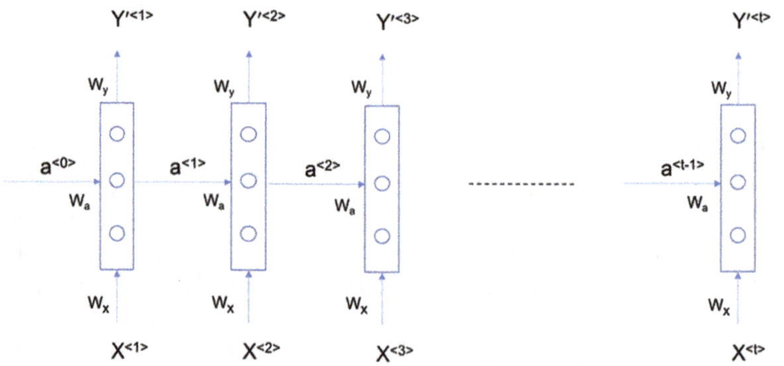

Fig. 11.2 Unfolded RNN structure

the state to the computation of the next output. The representation of RNN is presented in Fig. 11.2. The output of sequence t is represented in Eqs. 11.4 and 11.5:

$$a^{<t>} = g\left(W_a\, a^{<t-1>} + W_x X^{<t>} + b_a\right) \tag{11.4}$$

$$Y'^{<t>} = g\left(W_y a^{<t>} + b_y\right) \tag{11.5}$$

where

$a^{<t>}$ is computed activation input from sequence t
$X^{<t>}$ is input at sequence t
$Y'^{<t>}$ is output computed at sequence t
W_x is weight vector for input vector
W_y is weight vector for output vector
W_a is weight vector for activation input
g is activation function: *tanh* for activation input and *Sigmoid* for output

RNN representation in Fig. 11.2 shows the network computing state at each of the sequence inputs and output. From Eq. 11.5, it is clear that this is the same as a simple neural network except that the activation is based on the past and the present inputs and in some cases future inputs (non-real-time data, not covered in this analysis). The primary change is in the activation input as shown in Eq. 11.4 that is not only computed based on the current input and corresponding input weights but also on the past inputs and the weights assigned to past inputs. It is important to note that the weights W_x, W_y, W_a remain the same at each of the sequences. This reduces the number of variables to compute for any size of unfolding in RNN models. The initial activation $a^{<0>}$ is initialized to all 0s. This representation can be simplified as shown in Fig. 11.3. This encodes the recurrent aspect of the network and shows the impact of past sequence elements in the computation of the current sequence.

The interesting aspect of the network is that the input weights and the output weights for the network to compute Y' are the same for all the sequence elements. The weights computed based on the previous sequence element are encoded in the

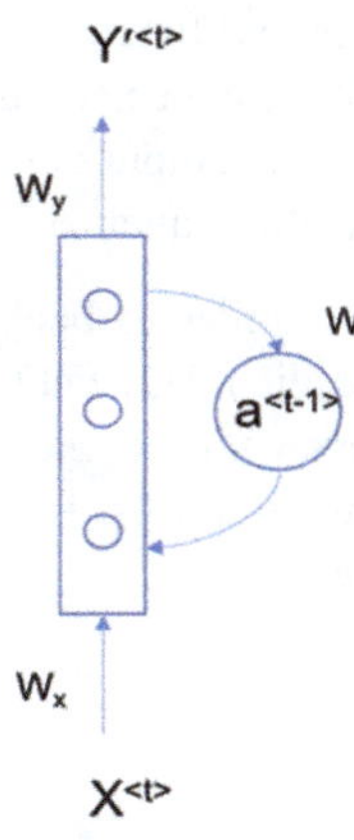

Fig. 11.3 RNN compact representation

activation of $a^{<t>}$ and are applied to the subsequent sequence computation. The diagram also represents the output at each of the sequence elements. The output at each of the sequence may not necessarily be meaningful – depending on the application. For example, if the network is being used to identify the sentiment, then the final output is the output with consequence for an entire given input sentence.

11.1.2 Backpropagation in RNN

The training phase of RNN computes W_x, W_y, W_a as shown in Fig. 11.3. The methods for computing these values follow a similar pattern as computing weights at different stages in a simple neural network. The error is computed at each stage of the network sequence, and the weights are adjusted in Gradient Descent fashion over the training set. Loss function can be defined as in Eq. 11.6:

$$L(Y', Y) = -Y^{<t>}\log\left(Y'^{<t>}\right) - \left(1 - Y^{<t>}\right)\log\left(1 - Y'^{<t>}\right) \tag{11.6}$$

where

L is loss function
$Y^{<t>}$ is expected output at sequence t
$Y'^{<t>}$ is output computed at sequence t
t is the sequence index

In RNN, the total loss function is simply the sum of loss at each sequence as we treat the response of the network as output at all the sequences. The sum of errors over the sequence is defined as in Eq. 11.7:

$$L(Y', Y) = \sum_{1}^{t} L^{<t>}\left(Y'^{<t>}, Y^{<t>}\right) \tag{11.7}$$

where

L is loss function
$Y^{<t>}$ is expected output at sequence t
$Y'^{<t>}$ is output computed at sequence t
t is the sequence index

Similar to backpropagation in *Neural Network* chapter, backpropagation uses Gradient Descent to adjust the weights during training. To compute the change in the loss with the change in the weights, taking a partial derivative with chain rule is shown in Eqs. 11.8a and 11.8b:

$$\frac{\partial L}{\partial W_y} = \frac{\partial L}{\partial Y'}\frac{\partial Y'}{\partial z}\frac{\partial z}{\partial W_y} \tag{11.8a}$$

$$\frac{\partial L}{\partial W_y} = (Y' - Y) * a^{<t>} \tag{11.8b}$$

where

L is loss at sequence t
Z is ($W_y\ a^{<t>}$) representation from Eq. 11.5
W_y is the weight vector for output computation
$Y^{<t>}$ is expected output at sequence t
$Y'^{<t>}$ is output computed at sequence t

The derivation of this equation is based on the partial derivative of Sigmoid function and product of partial derivative for a variable and a constant as covered in Neural Network chapter.

The primary thing to notice here is that the weight computation at each sequence for output weights depends only on the Y', Y and a at that sequence and hence can be computed with simple matrix multiplication.

Applying the same technique to backpropagate the error to compute weights W_a and W_x however is different as the derivative depends on the term for activation of the prior sequence. Eq. 11.9 shows the partial derivative chain rule for gradient for W_a:

$$\frac{\partial L}{\partial W_a} = \frac{\partial L}{\partial Y'} \frac{\partial Y'}{\partial a^{<t>}} \frac{\partial a^{<t>}}{\partial W_a} \tag{11.9}$$

where

L is the loss at sequence t

Notice that each of the terms at $a^{<t>}$ depends on $a^{<t-1>}$. Hence, computing terms for a requires computing values for all the sequences and using the previous error change terms in computing the current terms of vector a. This can be visually shown as presented in Fig. 11.4.

From Eq. 11.9 and Fig. 11.4, notice that the gradient gets smaller as the number of sequences increases if the initial error gradient is small. The gradients shrink exponentially fast if the initial error gradient is small. This makes it difficult to capture long-term dependencies of sequence elements. It is also possible that under some initialization conditions and activation functions, the initial derivatives are large that can lead to exploding gradient problem. This is usually dealt by having an upper threshold that limits the size of the gradient during propagation.

11.1.3 Vanishing Gradients

The sequence length in RNN can be large, and deriving context for long-range dependencies can introduce errors. For example, interactions between words that are from different parts of a sentence can change the meaning of the sentence. In the

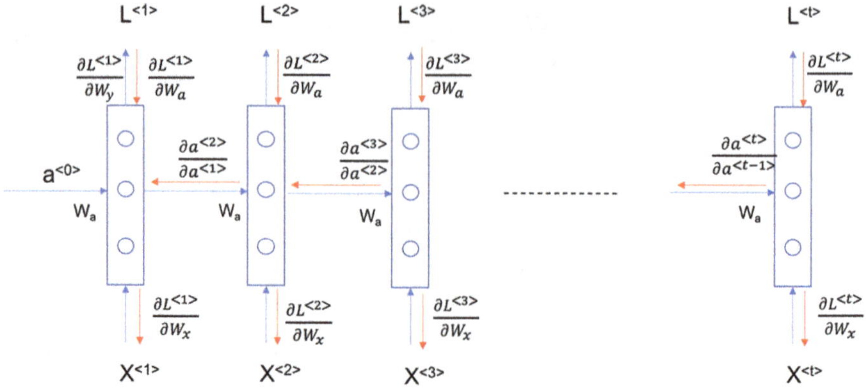

$$\frac{\partial L^{<2>}}{\partial W_a} = \frac{\partial L^{<2>}}{\partial a^{<2>}} \frac{\partial a^{<2>}}{\partial a^{<1>}} \frac{\partial a^{<1>}}{\partial W_a}$$

$$\frac{\partial L^{<2>}}{\partial W_x} = \frac{\partial L^{<2>}}{\partial a^{<2>}} \frac{\partial a^{<2>}}{\partial a^{<1>}} \frac{\partial a^{<1>}}{\partial W_x}$$

Fig. 11.4 Backpropagation through time for RNN

sentence "Chris lost balance when jumped to catch the ball and fell on his back," the pronoun *his* determines *Chris* as a short name for a male. The same sentence could have "her" as pronoun and will be entirely valid making *Chris* a short name for a female. Here, the context for *Chris* depends on the pronoun that occurs more than *ten* words apart in the sentence.

Computing the weights using derivatives of error change will get smaller as the sequence number increases. The effect of the pronoun in the sentence, in computing the propagation weights, gets vanishingly small. It is also possible that the gradient computation can be very large depending on the activation function and the weights and cause *NaN* computations for the gradients. A simple RNN will not be able to capture the long-term dependencies and will lead to failures in modeling.

Exploding gradients can be solved simply by capping the value to an upper bound or adjusted maximum. By restricting the maximum value, the propagation of the gradients can be controlled. Vanishing gradients can be addressed by implementing *ReLU* activation instead of *tanh* or *Sigmoid* function and initializing weights that are better suited for the context.

LSTM and *GRU* networks better address these problems and are addressed in Sects. 11.2 and 11.3.

11.2 LSTM

Long short-term memory (*LSTM*) networks are introduced to address the problem of long-term dependency modeling in RNNs. *LSTM* networks address the problem of vanishing gradients by designing a memory cell approach and providing gates that

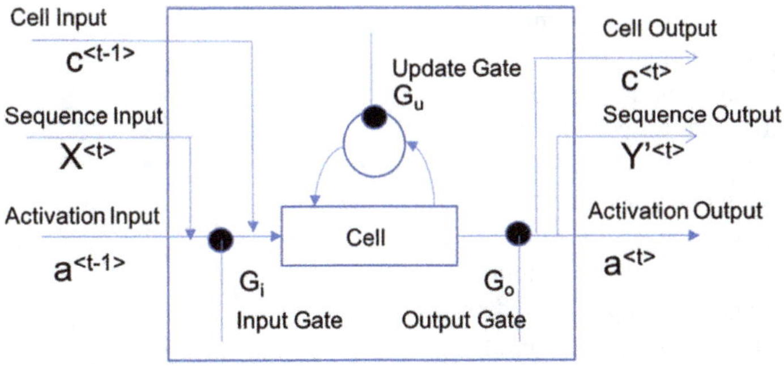

Fig. 11.5 LSTM model representation

control the influence of previous sequence on the current state and output of the current sequence. An intuitive model of this approach is shown in Fig. 11.5.

In Fig. 11.5, the input gate controls the effect of prior cells input on the current cell state. Update gate controls the degree to which the current sequence computation of RNN accounts for the cell state. Output gate controls the activation input of the current cell in predicting the output of the current sequence. The equations representing the application of the gate controls give better perspective of the usage of gate controls in the *LSTM* network as shown in Eqs. 11.10, 11.11, 11.12, 11.13, 11.14, 11.15, and 11.16:

$$Y'^{<t>} = g\left(W_y\, a^{<t>} + b_y\right) \tag{11.10}$$

Equation 11.10 is a familiar computation for neural network with activation function depending on the weights at the prior last layer and activation input.

$$a^{<t>} = G_o \tanh(c^{<t>}) \tag{11.11}$$

$$c^{<t>} = G_u\, c'^{<t>} + G_i\, c^{<t-1>} \tag{11.12}$$

$$c'^{<t>} = \tanh\left(W_c\, a^{<t-1>} + W_x X^{<t>} + b_c\right) \tag{11.13}$$

Equations 11.11, 11.2, and 11.13 show the change in activation $a^{<t>}$ computation based on the gates. The cell input using *tanh* activation based on the gates G_u and G_i are used for fine-tuning the memory and forget parameters in the current sequence output computation:

$$G_u{}^{<t>} = g\left(W_u\, a^{<t-1>} + W_x X^{<t>} + b_c\right) \tag{11.14}$$

$$G_i{}^{<t>} = g\left(W_i\, a^{<t-1>} + W_x X^{<t>} + b_c\right) \tag{11.15}$$

$$G_o{}^{<t>} = g\left(W_o\, a^{<t-1>} + W_x X^{<t>} + b_c\right) \tag{11.16}$$

where:

$a^{<t>}$ is computed activation input from sequence t
$X^{<t>}$ is input at sequence t
$Y'^{<t>}$ is output computed at sequence t
$c^{<t>}$ is cell output at sequence t
$c'^{<t>}$ is intermediate cell output at sequence t
W_x is weights for input vector
W_y is weights for output vector
W_c is weights for activation input
W_u is weights for computing update gate
W_i is weights for computing input gate
W_o is weights for computing output gate
g is activation function, tanh for activation input and Sigmoid for output

Long short-term memory (LSTM) networks have been very successful in modeling NLP problems with very good long-range dependency modeling. Language modeling for translating one language into another or suggesting the next word while typing and speech recognition are some of the applications of LSTM. Intuitively, memory cells control if the current sequence should have bearing on data occurring on the long term. There are multiple variations of LSTM networks that have shown promise in certain types of modeling. For example, peephole connection models include the previous cells output in the gate models instead of the activation state of the cells. One successful variation is a gated recurrent unit (GRU) implementation that simplifies the number of variables in LSTM model.

11.3 GRU

Gated recurrent unit (*GRU*) networks are introduced to simplify the LSTM network models with less number of variables to compute for generating the network. GRU models have been very successful in modeling NLP and are extensively used in RNN models. GRU models use relevance term and an *update* gate in place of input gate, *forget* gate, and output gate of LSTM network. An intuitive representation of GRU cell is presented in Fig. 11.6.

GRU network computes a relevance term and uses it to update the state of the cell. The *update* gate is similar to *forget* gate that modifies the output that is computed from the sequence to use long-term dependencies. Equations presented in 11.17, 11.18, 11.19, 11.20, 11.21, and 11.22 give better perspective on the effect of the gate terms in the prediction of sequence terms in the network:

$$Y'^{<t>} = g\left(W_y a^{<t>} + b_y\right) \tag{11.17}$$

$$a^{<t>} = c^{<t>} \tag{11.18}$$

$$c^{<t>} = G_f\, c'^{<t>} + \left(1 - G_f\right) c^{<t-1>} \tag{11.19}$$

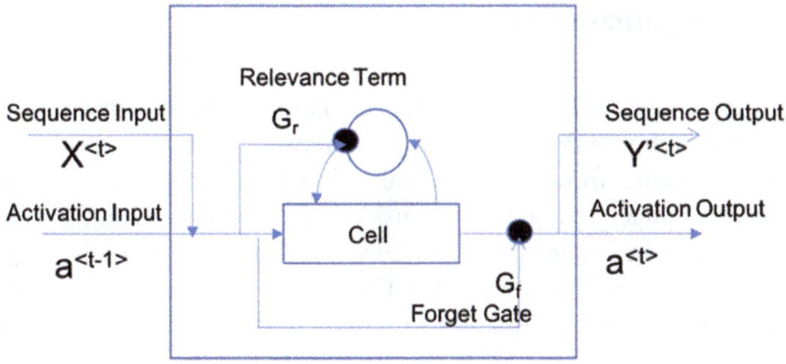

Fig. 11.6 GRP network cell representation

$$c'^{<t>} = \tanh\left(G_r * \left(W_c\, a^{<t-1>} + W_x X^{<t>}\right) + b_c\right) \tag{11.20}$$

$$G_r{}^{<t>} = g\left(W_r\, a^{<t-1>} + W_x X^{<t>} + b_r\right) \tag{11.21}$$

$$G_f{}^{<t>} = g\left(W_f\, a^{<t-1>} + W_x X^{<t>} + b_f\right) \tag{11.22}$$

where:

$a^{<t>}$ is computed activation input from sequence t
$X^{<t>}$ is input at sequence t
$Y'^{<t>}$ is output computed at sequence t
$c^{<t>}$ is cell output at sequence t
$c'^{<t>}$ is intermediate cell output at sequence t
W_x is weight for input vector
W_y is weight for output vector
W_r is weight for relevance vector
W_f is weight for computing forget gate
b is the regularization term
g is activation function: *tanh* for activation input and *Sigmoid* for output

One of the problems with *LSTM* and *GRU* networks is that they are sequential models and the parameters have to propagate sequentially for long-term dependencies. If the sequences are several terms away, for example, in DNA sequence modeling with repetitive patterns, then the models using *LSTM* networks are very slow and require trial and error. To improve on these models, a newer technique of *hierarchical neural attention encoder* is being considered. This technique creates a context vector based on the past sequence data that allows looking back into the past terms of sequences easier without having to pass through all the terms of the sequence. This is an evolving field and can expect significant advancements in performance in the future.

11.4 Self-Organizing Maps

Self-organizing maps (*SOMs*) are models of neural networks that are based on unsupervised learning. SOMs are feedforward networks with the network organization of neurons distinct from regular neural networks. SOMs are used in converting high-dimensional data into a lower-dimensional data with common features. SOMs are used in extracting features in high-dimensional input data similar to *principal component analysis* (*PCA*) (explained in Chap. 12); however, unlike PCA, SOMs can perform non-linear transformation, providing more accurate mapping than PCA in most cases.

SOM organization is similar to neuron organization in the brain, where the neurons that are related to sensory input are all spatially organized in close proximity to process the input with least distance transmission of signals during processing. Auditory sensors are packed closely together and are distinct from the visual sensor location in the brain. This organization is called *topographic maps*. This allows for efficient processing by keeping the information in contextual area. SOMs classify the input data simply by identifying the feature group mapping of the input in the network.

SOMs use *competitive learning* in which the input vector is used to choose a winner neuron among a set of neurons. The winner neuron is then activated and weights adjusted to organize closer to similar neurons in the network. This provides organizing neurons based on input features in unsupervised fashion as discussed in Sect. 11.4.1.

11.4.1 Representation and Training of SOM

The network for SOM is organized as a one-dimensional or two-dimensional lattice. Higher-dimensional spaces are possible; however, they are used less frequently. Each node or neuron on the lattice, usually on a rectangular or hexagonal grid, represents input vector with weights attached to each of the input vector dimensions.

Training the network consists of initializing, competing, and adapting connection weights to form self-organization. Initialization of the network can be performed by randomly assigning small weights or by performing PCA on the input space and evenly sampling two dominant eigenvectors to assign weights to the neurons.

The competitive process is to determine the winner by finding the lowest value for distance function or discriminant function. The discriminant function can be a simple distance vector in Euclidean space as shown in Eq. 11.23:

$$d_j(x) = \sum_{i=1}^{D} \left(x_i - w_{ji}\right)^2 \tag{11.23}$$

where

x is the input vector
d_j is the discriminant function for jth neuron
D is the dimension of input sequence
i is the member of D-dimensional input sequence
w_{ji} is the weight of the jth neuron corresponding to ith element of the input vector

Training the network consists of updating weights associated with the neurons based on the training input. The neurons in map space stay fixed, while the weights on input features are updated with each of the input vectors. The winner neuron gets updated the most, and surrounding neurons to the winner are updated with a dampened correction. The update to the winning neuron and the surrounding neurons is represented in Eq. 11.24:

$$\partial w_{ji} = T_{jI}.\left(x_i - w_{ji}\right) \tag{11.24}$$

where

∂w_{ji} is the adjustment
I is the index of the winning neuron
i is the index of the input vector element
T_{iI} is a Gaussian function that determines the rate at which the adjacent neurons are updated

$$T_{jI} = e^{\left(-S_{jI}^2/2\sigma^2\right)} \tag{11.25}$$

where

T_{jI} is the update applied to jth neuron in the network
S_{jI} is the lateral distance between neurons j and I
I is the winning neuron
j is the index of neuron in the network
σ is the width from the center in Gaussian bell curve to apply the updates

Equations 11.24 and 11.25 provide the update to neighboring neurons based on the distance from the winning neuron. The winning neuron has the T_{II} value as *1* with *Gaussian* functional decrease as lateral distance increases. Possible variations on the decrease function are *Mexican hat* function that penalizes near misses more than just that lateral distance.

To create convergence of the organization of neurons over time, there is a time-based decay in the adjustment function that needs to be applied. This ensures that the anomaly data in the input does not sway the learning significantly and also ensures that the network converges.

The learning rate adjustment and the neighborhood size adjustments are made to Eq. 11.25 to include learning over time or sequences. The updated equation is shown in Eq. 11.26:

$$\partial w_{ji} = \Delta_t \, T_{jI}.(x_i - w_{ji}) \tag{11.26}$$

$$T_{jI} = e^{\left(-S_{jI}^2/2\sigma_t{}^2\right)} \tag{11.26a}$$

$$\Delta_t = \Delta_0 e^{\left(-\frac{t}{t_1}\right)} \tag{11.26b}$$

$$\sigma_t = \sigma_0 \, e^{\left(-\frac{t}{t_2}\right)} \tag{11.26c}$$

where

∂w_{ji} is the adjustment
I is the index of the winning neuron
i is the index of the input vector element
T_{iI} is a Gaussian function that determines the rate at which the adjacent neurons are updated
Δ_t is the adjustment at time t
t_1 and t_2 are time constants that are positive to decay over time
Δ_0 and σ_0 are values initialized

Choosing the parameters carefully, the network organizes itself into groups of neuron structures that classify and process different types of input messages. The training phase usually requires the input sequences that are order of few hundred times the number of neurons in the network for convergence. Choosing the parameters incorrectly can lead to errors in the map, and the network may not converge.

Applications of SOM include the analysis of financial stability, behavioral pattern detection in time series data in data centers, and mapping of different patterns in atmospheric sciences and in genomic studies in biodiversity and ecology.

In this chapter, we discussed neural network adoption for specialized applications like natural language processing, DNA analysis, financial stability analysis, and feature recognition in noisy data. Applications are evolving with the increased amount of data and computational power available that solve current problems and show great promise in advancing accuracy.

Chapter 12
Principal Component Analysis

Principal component analysis (PCA) is a statistical process that allows reducing number of variables from a given dataset to a smaller set of variables that can be used in data analysis. The reduced set of variables retain the variance present in the original dataset. This is very useful in machine learning where the amount of data required for training is related to number of variables used in modeling.

Analysis and modeling require collecting data in multiple dimensions or ask set of questions that could bring out underlying patterns. This will mean that some of the dimensions or questions will have answers that do not add information to the dataset. Consider a set of variables in measurement that are correlated with each other. For a trivial example, given that total volume of water in a cup is known, we only need to know either volume of water in the cup or volume available to be filled with water, and the other variable can be derived knowing the total volume which is a constant. In this case, there is no need to include both the variables as they add no new information with new measurements. Now, consider variable representing volume of the cup. This is also superfluous as continuous measurement does not add additional information to the data.

In this simple example, given three dimensions of measurements, volume of the cup, volume of water in the cup, and volume that can be filled with water, PCA can reduce them to one variable that includes change in information with change in water levels.

PCA can be defined formally as a statistical procedure used to map a set of interrelated variables into a smaller set of linearly uncorrelated variables while retaining as much variance as possible in the original dataset.

G. Rebala et al., *An Introduction to Machine Learning*,
https://doi.org/10.1007/978-3-030-15729-6_12

12.1 Applications of PCA

12.1.1 Example 1

Suppose you are asked to build a predictive model of a stock performance. For accurate modeling, you would consider all the dimensions of measurements related to stock performance. These variables can span from different categories of financial ratios, labor market, housing variables, sentiment measures, GDP measurements, inflation, and unemployment along with closing price, opening price, intraday high, intraday low, alpha, beta, and fundamentals of individual stock. A quick analysis of variables provided by a major stock brokerage firm for selecting shares provides more than 30 variables for analysis.

Some of the variables may be interrelated, and some of them may not even add any additional information for the analysis. It may not be obvious even to financially literate person to intuitively evaluate the correlations among all the variables. PCA can be applied to these large set of variables and extract a reduced set of alternative variables that represent close to entire variance in the original data. The smaller set can then be used to evaluate different models for stock performance.

12.1.2 Example 2

Capacity planning in cloud environments is a complex problem. Increasing complexity of applications deployed in cloud environments with unpredictable virtual resource performance in shared environment makes capacity optimization challenging. In one of the studies, it was shown that on premise data centers use less than 20 percent of their available capacity. Furthermore, resource allocation is flexible through increase and decrease of resources. This flexibility makes modeling of performance to automate capacity optimization a continuous process.

Servers deployed in the cloud collect a large number of metrics that include network usage, memory usage, compute usage, and storage access metrics. Application services depend on the infrastructure as well as service performance for end user delivery. For network monitoring, the metrics collected include metrics for errors on packets, transmit errors, receive errors, bytes received, bytes transmitted, packets received, packets transmitted, header errors, and discarded packets for *ipv4* and *ipv6* protocols. One can see that there are many metrics and some of these metrics have high correlation with each other.

The number of metrics collected for a server deploying application code can collect more than 300 metrics. Modeling performance of an application based on all the collected metrics makes it difficult to interpret or create a model.

A typical server model can reduce the number of variables or *principal components* to model the performance to less than 20% of original number of variables by applying PCA.

Performance models using the reduced *principal components* can then be used to plan for capacity requirements, performance bottleneck detection, and performance prediction.

12.2 Computing PCA

Principal components are variables in alternate dimensions that can be substituted for original variables in the analysis. A simple solution to derive *principal components* using linear algebra is based on covariance matrix. A generic version of PCA can be derived using *singular value decomposition* (*SVD*) method. PCA works well under certain assumptions of properties of observed data. This section discusses the mathematical foundation of PCA.

12.2.1 Data Representation

Let X represent the observations of all the data and variables. Then X can be represented X_{mn} matrix where m represents the number of observations and n represents the number of variables being measured:

$$X = \begin{bmatrix} x_{11} & \cdots & x_{1n} \\ \vdots & \ddots & \vdots \\ x_{m1} & \cdots & x_{mn} \end{bmatrix}$$

12.2.2 Covariance Matrix

Computing relative relationship between variables can be done by using covariance computation. If there is no relationship between variables, then the covariance will be 0, and if there is high correlation between variables, then the covariance value will be close to 1 with positive correlation or -1 for negative correlation.

Let Y be a matrix with dimensions (m, n), and then the covariance matrix for Y is given by Eq. 12.1:

$$E = \frac{1}{m}\left(YY^{\mathrm{T}}\right) \tag{12.1}$$

where:

Y^{T} is transpose matrix of Y

12.2.3 Diagonal Matrix

A matrix is a diagonal matrix if all the elements of matrix are 0 expect for the diagonal elements. If the matrix E in Eq. 12.1 is a diagonal matrix, then the elements of Y are not correlated with each other and account for variance of all observations.

12.2.4 Eigenvector

Eigenvector is a vector that when applied to a square matrix M will only change the scalar part of the matrix and leaves the direction unchanged. This can be represented as Eq. 12.2:

$$MV = aV \tag{12.2}$$

where:

V is a vector, also the *eigenvector* of M
a is a scalar, also the *eigenvalue* corresponding to V

12.2.5 Symmetric Matrix

A symmetric matrix has the property that the transpose is equal to the original matrix, as represented by Eq. 12.3:

$$M = M^{\mathrm{T}} \tag{12.3}$$

12.2.6 Deriving Principal Components

For a given matrix X, if you can find a matrix P such that Eq. 12.4a is satisfied, then rows of P represent *principal components* of X:

$$Y = PX \tag{12.4a}$$

where:

(YY^{T}) is a diagonal matrix

Consider a theorem that states, for a symmetric matrix M, there exists a diagonal matrix D such that Eq. 12.4b is satisfied:

$$M = EDE^{\mathrm{T}} \tag{12.4b}$$

where:

D is a diagonal matrix
E is *eigenvectors* of M

Without actually deriving the result, applying the theorems corresponding to Eqs. 12.4a and 12.4b, it can be seen that if P is selected as *eigenvector* of X in equation $Y = PX$ then YY^{T} is a diagonal matrix.

Hence, the *principal components* of X are *eigenvectors* of XX^{T} which are rows of P.

12.2.7 Singular Value Decomposition (SVD)

A singular value decomposition is a generalization of above technique for any type of matrix. A singular value decomposition of a matrix is finding factors of matrix X such that a covariance matrix is a product of orthonormal vectors U, V, and diagonal matrix D with positive real entries, as shown in Eq. 12.5. It can be shown that columns of vector V represent *principal components* in this decomposition:

$$X = UDV^{\mathrm{T}} \tag{12.5}$$

An intuitive interpretation of the Eq. 12.5 is that the data points represented by m-dimensional space of X are transformed by n-dimensional space of V where $n < m$ and scaled by D to minimize the mean squared error of the resulting data points on to the plane of n dimensions.

12.3 Computing PCA

12.3.1 Data Characteristics

Principal components can be expressed as a linear equation of observed metrics. There is research on dealing with nonlinearity of the variable relationship called kernel PCA. This section primarily focuses on linear PCA.

Principal components derived using PCA account for overall *variance* of the original dataset in the variance of computed PCs with smaller number of variables. There is an implicit assumption that mean and variance describe the entire distribution of variables under observation. This zero mean distribution that can be described by variance is true only in *Gaussian* or normal distribution.

Large variance in observed variable will contribute most to the overall variance of computed principal component. If the observed values of variables have very different ranges, then the data needs to be normalized/scaled (Sect. 4.2) across the variables when applying correlation analysis for principal component computations.

12.3.2 Data Preprocessing

Consider the example of two variables measured being distance and time. If the distance is measured in centimeters and time in hours, then the resultant principal component is skewed toward the axis of distance. However, if the distance is measured in kilometers and the time in seconds, then the principal component computed will skew toward the axis of time instead of distance.

To overcome the problem of skew due to scale of variable value, Feature Scaling through standardization is performed as preprocessing step on the data. Standardization, also called *z-mean normalization*, converts the data to mean of 0 and standard deviation of 1. Equation 12.6 shows the method for *z-normalization*:

$$X_{ij} = \frac{X_{ij} - \mu}{\sigma} \tag{12.6}$$

where:

X_{ij} is data point at index i
μ is the mean of the dataset for variable X, corresponding to column j
σ is the standard deviation of the dataset for variable X, corresponding to column j

12.3.3 Selecting Principal Components

Principal components are computed from computing a covariance matrix and performing *SVD* on the resultant matrix (Sect. 12.2):

$$\text{Sigma} = \frac{1}{m}\left(Y\,Y^{\mathrm{T}}\right) \tag{12.7}$$

$$UDV^{\mathrm{T}} = SVD(\text{Sigma}) \tag{12.8}$$

In Eqs. 12.7, 12.8, and 12.9, the matrix U represents the *eigenvectors* of the matrix Y that is used to compute principal components:

$$\text{Principal components} = U^{\mathrm{T}} Y \tag{12.9}$$

Selecting k columns of matrix U will provide k principal components. The value of k can be selected based on the amount of variance one wants to capture from the principal components.

Assume k principal components are selected from the matrix U represented by $\boldsymbol{U_k}$. The original dimension data can be computed from the reduced principal components as per Eq. 12.10. This ability to recreate the original dimension data can also allow you to compute the error introduced due to data reduction, as given by Eq. 12.11. Equations 12.12 and 12.13 provide a way to measure the error in variance captured by k principal components:

$$Y = \boldsymbol{U_k} \cdot PC \tag{12.10}$$

where:

Y is the output or result matrix
$\boldsymbol{U_k}$ is matrix with k principal components selected from *eigenvector U*
PC is matrix of coefficients of computed *principal components*

$$E_p = \frac{1}{m} * \sum\nolimits_{i=1}^{m} \left(y_i - y_{\text{project } i}\right)^2 \tag{12.11}$$

where:

E_p is error in variance projected
y_i is the original value of y
$y_{\text{project } i}$ is the computed/reconstructed value with k principal components selected
m is the number of observations of y

$$v_t = \frac{1}{m} * \sum\nolimits_{i=1}^{m} (y_i)^2 \tag{12.12}$$

where:

v_t is variance of y
y_i is the original value of y
m is the number of observations of y

$$E_t = \frac{v_t}{E_p} \tag{12.13}$$

where:

E_t is error in variance captured by k principal components

E_p is error in variance projected
v_t is variance of y

Typically, you would like to have the error in variance captured to be less than 0.01 (i.e., 1%), and the components are selected to have the *principal component* capture 99% of the original variance in dataset.

From Eq. 12.8, the diagonal matrix D captures the variance attributed to each of the *principal components* in descending order. Hence, the error can easily be computed by simply computing the proportion of the total variance computed:

$$\text{Error} = 1 - \frac{\sum_i^k D_{ii}}{\sum_1^n D_{ii}} \tag{12.14}$$

where:

k is the number of *principal components* used
n is the total number of *principal components* computed
D is the diagonal matrix in SVD decomposition
i is the index of the cell in the matrix

12.4 PCA Applications

12.4.1 Image Compression

Image compression is a common application of PCA using dimension reduction technique. Figure 12.1 shows a picture of moon taken with resolution of 1200 × 795 pixels.

Performing PCA using *prcomp ()* function in *r*, you can compute standard deviations associated with *principal components* and the principal components. The corresponding code (in *r*) is given below:

Fig. 12.1 Original image before compression

```
moon <-- readJPEG("moon.jpg")
moon_pca <- prcomp(moon, center=FALSE)
percent_variance = sum((moon_pca$sdev[1:k])^2)/sum((moon_pca$sdev^2))
```

PCA results are now stored in *moon_pca* object with *moon_pca$stddev* containing the standard deviations associated with *principal components*, and *moon_pca$rotation* is the matrix of PCs. *k* in the last line of the code denotes the number of *principal components* being considered and, hence, impacts the percent variance. Table 12.1 shows the percent variance captured by *principal components*, depending on the value of *k* (i.e., the number of *principal components* considered).

Hence, by choosing just 10 PCs, you can recreate the image with 98% accuracy. This is a compression of 10/795.

Original data can be reproduced from PCs using the following code excerpt.

```
moon_compressed = moon_pca$x[,1:k] * transpose(moon_pca$rotation[,1:k])
```

Pictures reproduced using 5 PCs and 20 PCs are shown in Figs. 12.2 and 12.3, respectively.

Figure 12.4 represents variances associated with number of principal components.

12.4.2 Data Visualization

Datasets with large number of variables usually have a smaller subset of variables that capture most of the variance represented by the variables. One way to

Table 12.1 Percent variance for difference numbers of principal components

Number of PCs	Percent variance
5	96.3
10	98.1
20	98.9
100	99.9

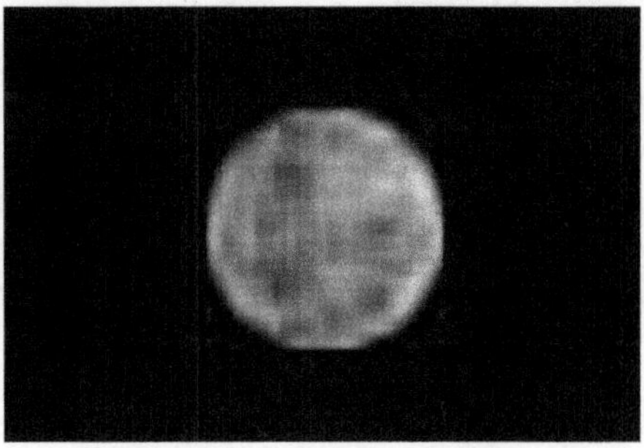

Fig. 12.2 Image reconstructed with 5 principal components

Fig. 12.3 Image reconstructed with 20 principal components

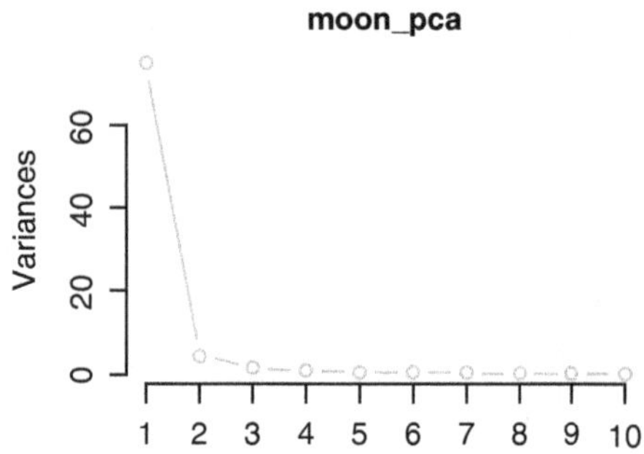

Fig. 12.4 Plot of variance captured with number of principal components

understand the application of PCA is to tease out the subset of variables that represent the data in a different subspace while preserving most of the variance of the original dataset.

Consider a two-dimensional dataset like base and height of buildings with data plotted in a plane. Applying PCA will tease out new dimension space that captures variance in the components.

Original dimension data is plotted in Fig. 12.5 and the alternate dimension plotted in Fig. 12.6. The original dimensions of x and y are transformed into alternate dimensions of *PC1* and *PC2*. In this example, you can see that most of the variance is captured by *PC1* in transformed dimension. The transformed data captures linear relationship between x and y in *PC1* and any differences in this linear relationship are captured in *PC2*.

Variable *PC1* may not have any physical meaning like *width* or *height* of the building; however, it is useful in visualizing and exploring the data which has significant benefit if a number of dimensions involved are large.

In the above example, the transformation can be interpreted as simple dimensional transformation of mapping data points in x, y coordinates with rotation to a coordinate system. The rotation of coordinates is equal to the slope of the line to attain principal coordinate with *PC1* and *PC2* matching new axis.

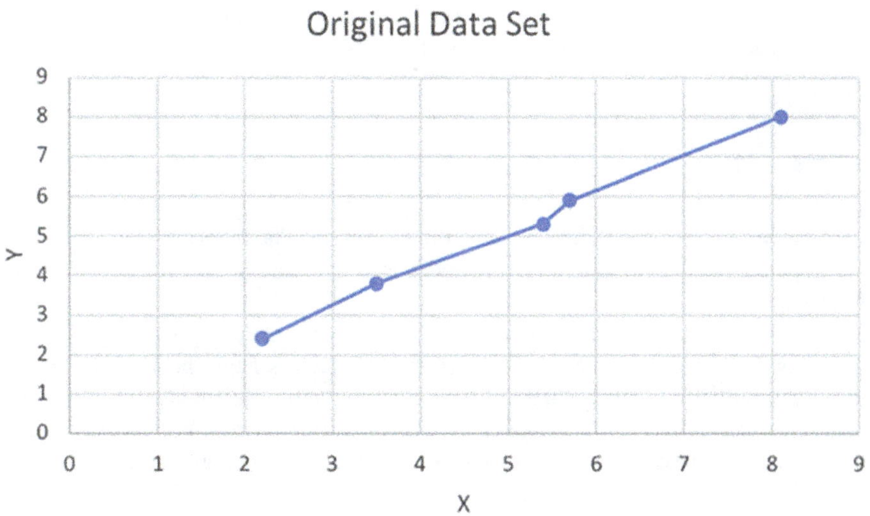

Fig. 12.5 Data visualization with principal components – original data

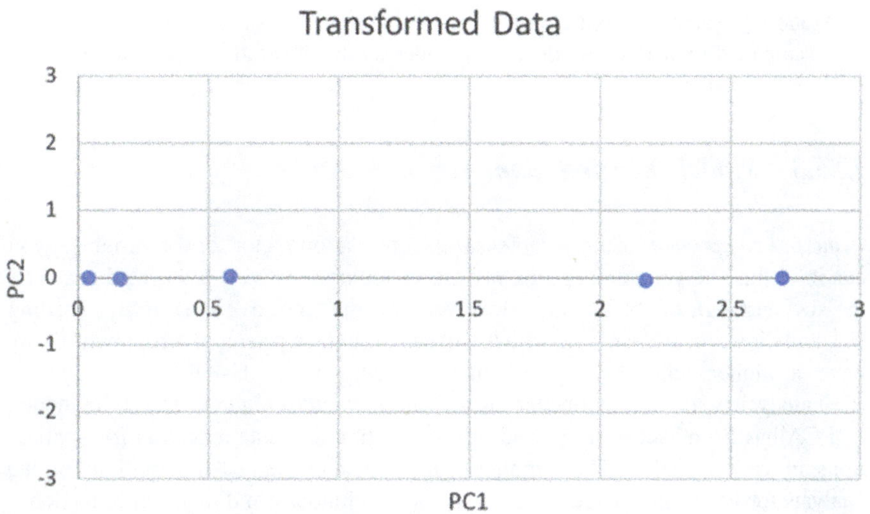

Fig. 12.6 Data visualization with principal components – transformed data

12.5 Pitfalls of PCA Application

12.5.1 *Overfitting*

Model construction overfitting occurs when the number of features is large and sufficient data is not there. To reduce number of features, sometimes PCA is used under the assumption that the reduced number of features will avoid overfitting.

Using PCA to avoid overfitting can lead to bad results. The problem of overfitting is better addressed using regularization techniques and better datasets.

12.5.2 Model Generation

PCA is used to generate PCs as reduced dimension set. The reduced dimension data is computed and then used to train the model using Logistic Regression to model the system.

The use of PCs to construct model can cause large errors in predicting data for new inputs. One alternative is to extract original factors from the data. *Common factor analysis* is a method that identifies combination of original variables that can maximize variance of observed variables. This is different from PCA which maximizes total variance of original variables. These are also called *latent factors* or hidden variables that encapsulate combined variance of observed variables. This method of computing factors is useful when the correlation matrix has high values not only in the diagonal but also other parts of the matrix. The latent factors tend to perform better in modeling than PCs.

Modeling based on original variables should be preferred, and only if issues arise with scale of dimensions should you consider model generation based on PCs.

12.5.3 Model Interpretation

Principal component loadings represent the correlation factor for the variables; you can infer that the variable is positively associated or negatively associated based on the size and sign of the loading. However, you must not overinterpret the loading size and direction as a variable with low loading value in principal component 1 may have a higher values in other principal components. Using *common factor* (CF) analysis will provide a better interpretation when trying to interpret the model.

PCA has been successfully used in visualization and data reduction for applications in various fields. The simplicity and the mathematical foundation for the analysis for PCA have made it one of the best techniques in dimension reduction.

Chapter 13
Anomaly Detection

The learnings in this chapter will help you determine if something is anomalous. Anomalous in this context effectively means out of range. It could be about a defect in a component or a fraudulent exchange. For example, if a person usually types with a certain speed, and suddenly the system sees a different speed, it is anomalous behavior, giving an indication of possible impersonation. Or, for a component, certain measured parameters being outside the range compared to the range exhibited by normal components could indicate a defective component. This information can be used for quality checks, either during manufacturing, shipment, acceptance stages, or also during preventive maintenance. The concept of anomaly detection depends on being able to observe certain parameters and then being able to form an opinion about some dependent behavior, which is difficult to directly observe at this time. For example, predict the expected remaining life of an aircraft engine based on noise (something that can be observed now). Similarly, use measured value of typing speed to predict if the person (not visible) at the other end is impersonating. Anomaly detection is used very extensively in monitoring computer equipment in large data centers. All the computers in a data center are monitored for many parameters. If a specific computer seems to exhibit a behavior that is very different from other computers, this computer could be defective and may need to be taken out of the network. For example, if it's getting too few jobs compared to other computers, maybe, it's working slowly. Or, if it's getting too many jobs compared to others, maybe, it's not working properly and terminating jobs immediately that come to it and, thus, be ready to take on additional jobs.

G. Rebala et al., *An Introduction to Machine Learning*,
https://doi.org/10.1007/978-3-030-15729-6_13

13.1 Anomaly vs Classification

Section 3.5 explained the concept of Logistic Regression, which gives a *yes/no*, *true/false* classification-type answer. In case of anomaly detection also, you are effectively dealing with a binary decision: normal or anomalous. So, it is pertinent to question the need for a separate model or mechanism for anomaly detection.

Logistic Regression depends on a large number of training data for both kinds of results: true/false. Often, you may not have a large number of data for anomalous behavior. You may have a huge data, but, mostly for normal situations. This lack of a large number of data corresponding to anomalous behavior motivates the need for a mechanism that would work for learning algorithms, where the available dataset is small.

Anomaly detection depends on the ability to predict normal behavior. It can be an anomaly if it occurs out of order, but individual value may be normal. For example, in a sequence 1, 1, 2, 3, 5, 8, 13, 20, 33, 53, number 20 is an anomaly as it is not the sum of previous two numbers. Hence, the models to detect anomaly need to be specialized for a given situation. Another example is that if a website receives more traffic during the day and less during the night and the traffic slowly tapers off, then a sudden increase in traffic in the evening even though it is within the acceptable range for the day will be an anomaly. These anomalies are determined using time series modeling.

Now that you understand the need, as you progress through this chapter, the models for different situations are addressed.

13.2 Model

Figure 13.1 explains the concept of model used for anomaly detection. Assume that a behavior may be identified as normal vs anomalous, based on just two attributes. The various marks (x) in the figure represent normal behavior. Note that we are not talking about the existence of data for the anomalous behavior.

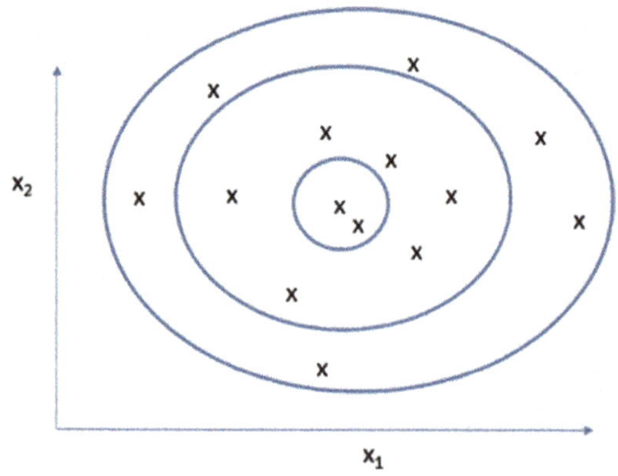

Fig. 13.1 Model for anomaly detection

The model should be such that the innermost value will have a very high value, and as you go outward, the values will decrease. So, you need to find a model $p(x)$ and a value ϵ such that:

$p(x) < \epsilon$ means anomalous behavior
$p(x) \geq \epsilon$ means normal behavior

where x represents $x_1, x_2, x_3, \ldots$, etc., the various features corresponding to the observed data.

Using the example of computers in data centers, the values on either side (exceptionally higher or lower number of jobs) should result in lower value of $p(x)$.

13.2.1 Distribution Density

Each of the features $x_1, x_2, x_3, \ldots$, etc. are modeled as a *normal* distribution. While formula corresponding to *normal* distribution might appear a bit daunting, this is one of the most popular statistical distribution model, meaning that most software packages should have pre-built libraries available for use. So, you only have to understand the concepts, rather than exactly knowing how to compute the various values. For the sake of completeness, though, we are providing the formula in this chapter.

For a variable X that follows a *normal* distribution, it is represented as shown in Eq. 13.1:

$$X \approx N\left(\mu,\ \sigma^2\right) \tag{13.1}$$

where:

$\approx$ represents belongs to
N represents a normal distribution
μ and σ^2 represent parameters that characterize the distribution

In order to avoid confusion between two different uses of the word "parameter," in this chapter, henceforth, we will use:

"parameter" to mean μ and σ^2 that characterizes a normal distribution
"features" to mean properties of a normal or anomalous data that are being monitored

It would be good to know that the parameters μ and σ^2 are sufficient to fully describe a *normal* distribution. For such a *normal* distribution, the probability of any value x is given by Eq. 13.2:

$$p(x) = \frac{1}{\sqrt{2\pi}} e^{-\left(\frac{x-\mu}{2\sigma^2}\right)} \tag{13.2}$$

The relations in Eqs. 13.1 and 13.2 are often represented in short simply by expression, $p(x, \mu, \sigma^2)$, which means probability of x in a *normal* distribution that is characterized by μ and σ^2. Refer to Sect. 1.4.5 for discussion on normal distribution.

13.2.2 Estimating Distribution Parameters

Consider you have m training data, x^1, x^2, x^3, ..., x^m. Each training data has n features, $x_1, x_2, x_3, \ldots, x_n$. For now, assume that each of the training data is normal data. Even if a few of these data correspond to anomalous behavior, it is alright to still consider this data as normal for the purpose of estimating the parameters.

For each of the features, $x_1, x_2, x_3, \ldots, x_n$, determine the mean ($\mu$ – using Eq. 13.3) and variance (σ^2 – using Eq. 13.4) for that feature. Thus effectively, you are modeling each attribute individually as a normal distribution with its own parameters:

$$\mu = \frac{1}{m} \sum\nolimits_{i=1}^{m} x^i \tag{13.3}$$

$$\sigma^2 = \frac{1}{m} \sum\nolimits_{i=1}^{m} \left(x^i - \mu\right)^2 \tag{13.4}$$

Thus, applying Eqs. 13.3 and 13.4, over each of the n features, you now have values of μ and σ^2 for each of these n features, which can be denoted as $\mu_1, \mu_2, \mu_3, \ldots,$ μ_n and ${\sigma_1}^2$, ${\sigma_2}^2$, ${\sigma_3}^2$, ..., ${\sigma_n}^2$. So, you now have the normal distribution that each feature follows.

13.2.3 Metric Value

For any data x, you can determine the probability of each feature using the notation $p(x, \mu, \sigma^2)$ explained in Sect. 13.2.1. And, the final metric is given by the product of each of these probabilities. This final metric is represented mathematically by Eq. 13.5:

$$p(x) = \prod\nolimits_{j=1}^{n} p\left(x_j, \mu_j, {\sigma_j}^2\right) \tag{13.5}$$

where:

$\prod_{j=1}^{n}$ represents product of 1st value through nth value

$p(x_j, \mu_j, \sigma_j^2)$ represents probability of *j*th feature for a *normal* distribution with parameters μ_j and σ_j^2

The $p(x)$ given by Eqs. 13.2 and 13.5 is conceptually different. Equation 13.2 gives the probability for a single feature, while, Eq. 13.5 gives the total probability for any given data, considering all the features. You can even combine these two equations to get the final metric value, as given by Eq. 13.6:

$$p(x) = \prod_{j=1}^{n} \frac{1}{\sqrt{2\pi}} e^{-\frac{(x_i-\mu_j)^2}{2\sigma_j^2}} \tag{13.6}$$

13.2.4 Finding ϵ

You now need to determine a suitable value of ϵ such that $p(x)$ as given by Eq. 13.5 (or 13.6) exceeds ϵ for a normal data and is below ϵ for anomalous data. This value of ϵ is the *threshold* – a boundary segregating normal data and anomalous data.

Use a set of training data to obtain the parameters μ and σ^2 for each feature. Now, use a set of labelled data (say, cross validation set) to determine $p(x)$ for each of these data. This cross validation set should have both normal data and anomalous data.

Considering each of the normal data, find the one that provides the minimal $p(x)$. ϵ value is so chosen that it should be lower than this minimal value of $p(x)$. This provides an upper bound for the value of ϵ.

Now, considering each of the anomalous data, find the one that provides the maximum value of $p(x)$. ϵ value is so chosen that it should be higher than this maximum value of $p(x)$. This provides a lower bound for the value of ϵ.

You can now choose an ϵ that satisfies both these criterion.

What if it is impossible to satisfy both these criterion simultaneously, i.e., the lower bound is higher than the upper bound? This indicates that the choice of features is not correct (besides, of course, possibility of an error in implementing the algorithm). Section 13.2.5 explains the course of action in such cases.

13.2.5 Validating and Tuning the Model

You now apply your model on a set of test data, having a combination of both normal data and anomalous data. If you find that the model does not work properly with the test data, it indicates a need for having a relook at the set of features. Look at some data that gives incorrect result, and see if it indicates certain features that should have been considered and are not considered. For example, say, an anomalous data in the test set was identified as normal data. Now, look at the features of this incorrectly identified data and a normal data, and try to find which all features of these data differ significantly. And, confirm that each of the features is considered in your modeling of $p(x)$.

The other thing you should check is that each of these features shows a *normal* distribution. The whole model is based on the assumption that each of the features is individually following a *normal* distribution, and that should be ensured. If the feature does not follow a *normal* distribution, you should apply certain mathematical operations to raw data so that the transformed data is closer to *normal* distribution, as explained in Sect. 8.9.

13.3 Multivariate Gaussian Distribution

Gaussian is another (and more mathematical) name for *normal* distribution. So, this section is about interaction of multiple variables (multivariate).

Sometimes, each of the features independently might look normal; however, their ratio might be of importance. Consider the data shown in Fig. 13.2.

xs in the figure represent normal behavior, and the circle represents an anomalous data. Considering the anomalous data, each of the two features x_1 and x_2 independently seems to be within their respective ranges. However, the ratio x_1/x_2 is high for this anomalous data. This kind of anomaly may not be caught by Eq. 13.5, which considers each feature individually and looks at their product. Rather, the need is for a model which considers all the features together, where their interrelation can also be considered.

The metric for such multivariate Gaussian distribution for data x_i is given by Eq. 13.7:

$$p\left(x_i, \mu, \sum\right) = \frac{1}{(2\pi)^{n/2}} \cdot \frac{1}{\left|\sum\right|} \cdot e^{-\frac{1}{2}(x-\mu)\sum^{-1}(x-\mu)^{\mathrm{T}}} \tag{13.7}$$

Since Eq. 13.7 seems relatively complex, we will understand it in a bit more details. Consider a dataset comprising of m data and each data having n features.

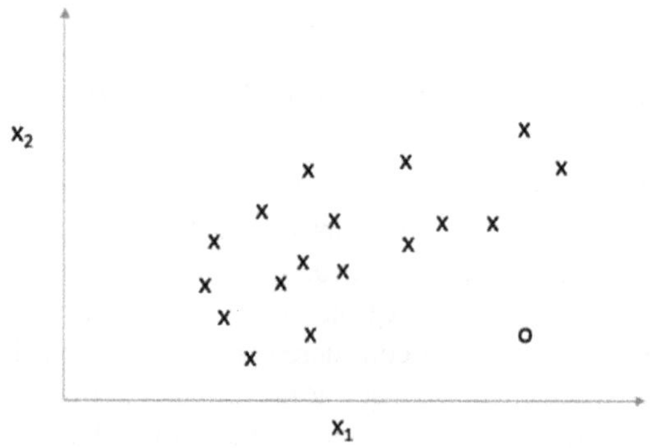

Fig. 13.2 Anomaly seen in ratio

This data can be represented as a matrix X having m rows and n columns, as shown below.

$$\begin{bmatrix} x_1^1 & \cdots & x_n^1 \\ \vdots & \ddots & \vdots \\ x_1^m & \cdots & x_n^m \end{bmatrix}$$

13.3.1 Determining Feature Mean

For each column (i) of this matrix, consider the mean (μ_i) for that column, using Eq. 13.3. These mean values may be thought of as a matrix of means, having a single row and n columns.

13.3.2 Determining Covariance

Determine a matrix $(X - \mu)$, as shown below. This matrix will also have a size of m rows by n columns, and its transpose matrix will have a size of n rows by m columns.

$$\begin{bmatrix} (x_1^1 - \mu_1) & \cdots & (x_n^1 - \mu_n) \\ \vdots & \ddots & \vdots \\ (x_1^m - \mu_1) & \cdots & (x_n^m - \mu_n) \end{bmatrix}$$

Determine covariance matrix $\sum$ using the Eq. 13.8:

$$\sum = \frac{1}{m}(X - \mu)^{\mathrm{T}}(X - \mu) \tag{13.8}$$

This covariance matrix has a dimension of n rows by n columns. The physical significance of this covariance matrix is that the off-diagonal elements of this covariance matrix denote the correlation between two features. So, element $\sum_{12}$ denotes the relation between features x_1 and x_2. A positive value means positive correlation, i.e., when one increases the other also increases and vice versa. Similarly, a negative value means negative correlation, i.e., when one increases the other decreases and vice versa. For a covariance matrix, its transpose will be exactly same as itself, i.e., $\sum_{12} = \sum_{21,}$ because the relation between features x_1 and x_2 will have same correlation, whichever way measured.

The inverse of this covariance matrix (expressed as $\sum^{-1}$) is used in Eq. 13.7 in the exponent portion. And, the determinant of this covariance matrix (expressed as $|\sum|$) is used in the denominator of Eq. 13.7.

13.3.3 Computing and Applying the Metric

Using the training data, you would determine the mean(s) for each feature, μ_1 through μ_n, as explained in Sect. 13.3.1. You would then determine the covariance matrix $\sum$ as explained in Sect. 13.3.2 and also compute its inverse and determinant.

Consider any data x^i. It will have n features, represented as $x_1^i, x_2^i, \ldots, x_n^i$, thus, effectively a row matrix with n columns. You can compute $(x^i - \mu)$, which again would be a row matrix with n columns. And, its transpose is a column matrix with n rows.

Looking at the exponent portion of Eq. 13.7, you have the following matrices being multiplied in order:

- $(x^i - \mu)$: *1 by n*
- $\sum^{-1}$: *n by n*
- $(x^i - \mu)^T$: *n by 1*

Thus, the final exponent value is a *1 by 1* – single number. Apply this value in Eq. 13.7, to obtain the value of the parameter p, and this can be compared with ϵ to determine if this data x^i corresponds to normal data or anomalous data. You can very easily establish that when all the features are independent, Eq. 13.7 reduces to Eq. 13.6.

13.4 Anomalies in Time Series

Time series data is a sequence of data points with associated time order or timestamp. The time sequence is usually equally spaced in time. In time series context, an anomaly is data that has low probability of occurrence based on the past data values ordered as a sequence.

Share price of a company in stock market changes often, and the representation of the share price with respect to time is time series data of the company share price. If the price changes suddenly during the day (even if the start and end price are within the expected range), it would be considered an anomaly.

Time series data is extensively used in monitoring systems including *Internet of Things* (*IoT*), stock exchanges, software applications and infrastructure, data centers, and in every day usage in trend analysis for geological events, population growth, and demographic changes. In computer networks, increase in traffic can be because of a *denial of service* (DoS) attack. In a stock exchange, drop in volume of trades could be because of an error in network. These issues can cause serious financial implications and hence require immediate detection (and, intervention).

Probability distribution techniques based on univariate analysis as in Sect. 13.2 or multivariate analysis in Sect. 13.3 are not sufficient to find anomalies in time series. Time series data sequence values change with time, and the distribution of values today may not be same as yesterday and can be completely normal for the system under consideration. A time series usually consists of a periodic element, a trend, and a noise element to the data. Finding anomaly in time series needs to take into account all the three variables.

13.4.1 Time Series Decomposition

A time series can be decomposed into a trend component, a seasonal component, and a random component that constitute the value of series at any given time. The time series as a composition of these three components can be expressed either by Eq. 13.9 or by Eq. 13.10:

$$Y_t = \text{Trend} + \text{Random} + \text{Seasonal} \tag{13.9}$$

$$Y_t = \text{Trend} * \text{Random} * \text{Seasonal} \tag{13.10}$$

Anomaly can be detected in a time series by adjusting for trend and seasonality of time series and finding the distribution of error component probability. Figure 13.3 shows an example data and then broken down into seasonal component, overall trend, and the residual, which may be considered as random fluctuation. After the decomposition, each of the components can be modeled independently and anomaly can be detected by estimating the probability of occurrence of a value at a given sequence or time.

13.4.2 Time Series Anomaly Types

Anomalies in time series can occur in many forms. Time series anomalies do not follow a general pattern that can be applied to all time series data. Anomalies are defined based on application and hence anomaly detection will depend on the application. However, there are common patterns that exist which can be used for defining anomalies. Time series data is generally composed of periodic data, a trend, and noise component as shown in the example of Fig. 13.4.

Figure 13.4 shows the temperature measurements in Berkeley, CA, over a period of 5 years. You can see the periodic nature of the seasonal temperatures and also a noise component as the data is not a smooth curve. The year 2015 has seen higher temperatures than other years, hence an anomaly for summer temperature.

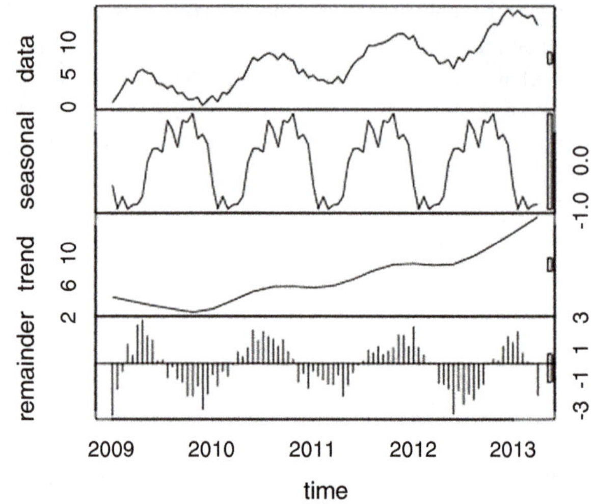

Fig. 13.3 Decomposition of time series data

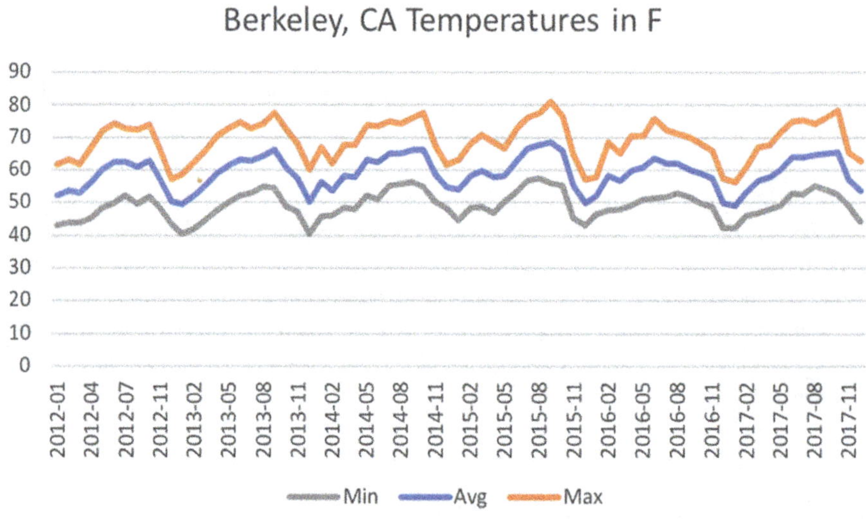

Fig. 13.4 Periodic nature of time series data

Anomaly in time series data with level change can be detected with a threshold-based system as in Fig. 13.5. The dotted line represents the threshold below which the data will be considered normal for the time series. Figure 13.6 shows linearly increasing trend in data with data varying around the trend. In this time series, if you account for trend as linearly increasing data with constant rate, the dotted line represents threshold to account for the increasing trend in the data. Anomaly detection then becomes a threshold detection problem while adjusting for trend as shown in Fig. 13.6.

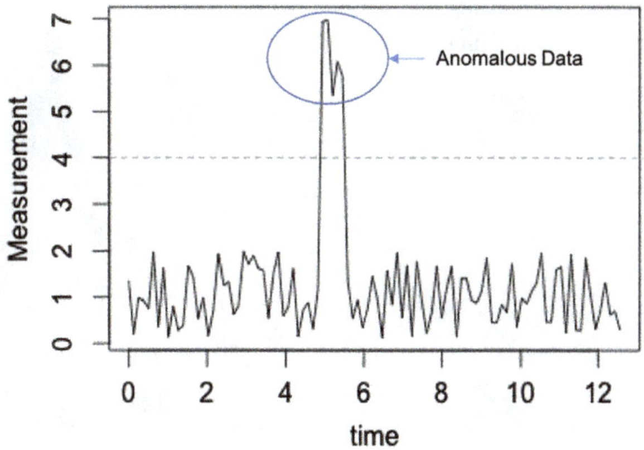

Fig. 13.5 Threshold violation in time series data

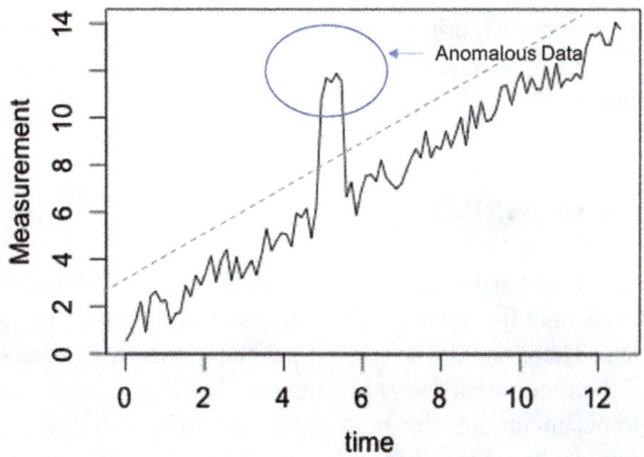

Fig. 13.6 Threshold violation in time series data with linear trend

Anomaly in time series can also occur because of change in periodic nature of the data or failure of expected levels in the series even though the levels are acceptable at a different time. In Fig. 13.7, anomaly occurs as the data does not reach the expected value at the time sequence.

13.4.3 *Anomaly Detection in Time Series*

The anomaly detection methods covered in Sect. 13.3 in utilizing Gaussian tails, probability estimations, and thresholds are also applicable to time series data. For example, Fig. 13.7 can be considered a data distribution where you can compute the distribution pattern and distribution density and apply the probability estimation of occurrence of a value to determine anomaly.

As you have seen in the examples, thresholds and probability estimation based on distribution densities are not sufficient to capture a large section of anomalies in time

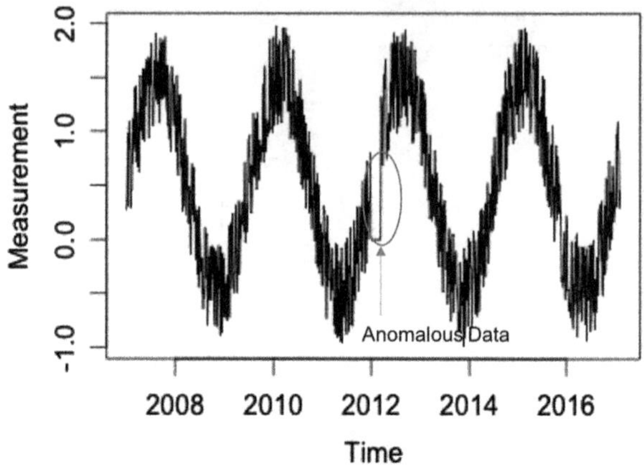

Fig. 13.7 Violation of expected pattern in time series data

series data. Algorithms like *ARIMA* and *RNN* are some of the methods that decompose the time series data into trend, seasonal, and random components and model the structures to find anomalies in periodic time series data.

13.4.3.1 ARIMA

Autoregressive integrated moving average (*ARIMA*) is used in modeling a time series data for understanding the time series better or for forecasting based on past data. There are multiple models that can forecast a time series data based on the type of the time series data. For example, *moving averages* is the average of past few data points to predict the next data point. *Exponential smoothing* is the variation of moving averages that reduces effect of values from distant past compared to immediate past. *Random walk* uses the current value and probability estimate for next value based on past errors in prediction. *Seasonal adjustment* is used to adjust for seasonal pattern by modeling within one season of the data series.

ARIMA is a generalization of all these models in a systematic way in identifying the time series properties and applying the models in a combined form. *ARIMA* works by linear combination of past values along with current and past values of error terms by using a systematic way of identifying the combination of past terms, their coefficients, and error adjustments needed in fitting a time series.

ARIMA model is based on three components that represent autoregression component, trend removal component, and moving average adjustment:

- The first p terms in the model represent the *autoregressive* (*AR*) component that accounts for correlation of previous values with current value.
- The next q terms represent the *moving average* (*MA*) for error terms after removing correlation for values that have occurred in the immediate past.
- The integrated (I) part is taking difference of time series terms to remove trend. This difference is represented by d. The differencing removes trend and seasonality in a time series and converts it to a stationary model making the series with a

constant *mean* and *variance* over time. In a stationary model, the probability distribution of values does not change with shift in time.

When $d = 1$, the differencing will replace Y_t with $Y_t - Y_{t-1}$. When $d = 2$, the differencing will replace Y_t with $Y_t - Y_{t-1} - (Y_{t-1} - Y_{t-2})$. Higher order of differencing maybe required for series with a trend that is changing with time. Second order of differencing is represented in Eq. 13.11. The optimal order of differencing is the order of differencing at which the standard deviation is the lowest for residual data:

$$y_t = (Y_t - Y_{t-1}) - (Y_{t-1} - Y_{t-2}) \tag{13.11}$$

For a series with no long-term trend, $d = 0$, as there is no adjustment required to make the series stationary. After converting the series into stationary series by differencing, any of the remaining autocorrelation in the time series terms needs to be compensated for model fitting.

Now, consider a time series where the current value only depends on the previous value of the series. A linear combination of past value with a coefficient and error can be represented by Eq. 13.12:

$$y_t = \phi_1 y_{t-1} + \epsilon_t \tag{13.12}$$

where:

ϕ_1 is the coefficient of the past value
ϵ_t is an error term compensating in the model

The value of y_{t-1} can be represented based on y_{t-2}. Substituting for y_{t-1} gives the Eq. 13.13. Expanding the terms gives Eq. 13.14:

$$y_t = \phi_2(\phi_1 y_{t-2} + \epsilon_{t-1}) + \epsilon_t \tag{13.13}$$

$$y_t = \phi_2\phi_1 y_{t-2} + \phi_2\epsilon_{t-1} + \epsilon_t \tag{13.14}$$

Now, extend the above model so that the current value depends on past p values of the time series, i.e., a lag of p. The error terms are not dependent on each other at different time intervals. The error terms need to be computed for each of the intervals sequentially. To account for decreased effect of past values further away from current time interval, the coefficient ϕ is exponentially smaller with increasing lag or further away from current time interval. Error terms on values further away from current time period have different effect on current value and are not related to the coefficient ϕ. To account for the change, introduce θ to adjust past errors on the current value. The model can be represented by Eq. 13.15:

$$Y_t = \phi_1 y_{t-1} + \phi_2 y_{t-2} \dots + \phi_p y_{t-p} + \epsilon_t + \theta_1 \epsilon_{t-1} + \theta_2 \epsilon_{t-2} + \dots + \theta_q \epsilon_{t-q} \quad (13.15)$$

where:

y_t is the time series value
ϕ are coefficients for past time series values for autoregressive compensation
θ are coefficients for error terms for moving average compensation
ϵ are identically distributed error terms with zero mean
p is the number of past values used to predict the current value
q is the number of past error terms to consider

For a stationary time series, autocorrelation in the time series terms are compensated for model fitting by choosing appropriate values for *autoregressive* (p) and *moving average* (q) terms. The values of p and q can be determined using plots of *Autocorrelation Function* (*ACF*) and *partial ACF* (*PACF*).

Autocorrelation Function (*ACF*) is a measure of similarity of time series with delayed version of itself. *Partial ACF* (*PACF*) is a measure of similarity with one of the terms when their mutual correlation is not explained by grouping with other terms. *PACF* adjusts the long-term correlation values by removing the short-term correlation values. Using *PACF* you can compute the correlation for different lags, say 1 through 6, and select the largest lag value that has significant correlation. This gives p. Using *ACF*, the largest value with high correlation will be q.

ARIMA model can be represented by (p,d,q) where p represents the *AR* component value, d represents the difference level, and q represents *MA* component.

Applying forecasting to a series with differencing and incorporating p and q computed based on *ACF* and *PACF* values is given by Eq. 13.16:

$$Y_t = \mu + \phi_1 y_{t-1} + \phi_2 y_{t-2} \dots + \phi_p y_{t-p} + \epsilon_t + \theta_1 \epsilon_{t-1} + \theta_2 \epsilon_{t-2} + \dots + \theta_q \epsilon_{t-q} \quad (13.16)$$

where:

y_t is the differenced time series value
μ is the mean of the series with differenced values
ϕ are coefficients for past time series values for autoregressive compensation
θ are coefficients for error terms for moving average compensation
ϵ are identically distributed error terms with zero mean
p is the number of past values used to predict the current value
q is the number of past error terms to consider

Forecasting of time series value at time t is computed by computing Y_t. The simplicity in finding the model parameters makes it attractive in modeling time series. *ARIMA* has been successfully applied to various time series forecasting including share price, agricultural output, and disease spread in populations.

13.4.3.2 Machine Learning Models

Neural networks are increasingly being employed to model time series. Deep neural networks like *TFLearn deep neural network* (*DNN*) with complete feedforward network are shown to successfully model the time series in modeling the traffic patterns of websites.

Recurrent neural networks (*RNNs*) (explained in Chap. 11) contain recurrent loops with feedback mechanism for errors to the input. *RNN*s with certain number of blocks for recurrent loops are effective in modeling sequence data with the seasonal components of time series as well as temporal state as short-term memory. These are found to perform as well as *DNN*s for modeling traffic patterns in websites.

Long short-term memory (*LSTM*) based on *recurrent neural networks* (*RNNs*) makes use of sequence dependence among the input variable. *LSTM* networks can employ complex structures that can compute the sequence of input along with long-term and short-term patterns in the sequence of input variables. *LSTM*s with their added complexity in constructing complex cell architecture do not perform significantly better than *DNN*s.

Neural network structures require optimization for constructing number of hidden layers and cell structure for seasonal components. *DNN*s and *RNN*s show significant promise in time series modeling and is an area of active research.

Chapter 14
Recommender Systems

This chapter explains how to build recommender systems. Recommender systems are the systems that make recommendations to you, based on prior choices exercised by you and by others who have exhibited similar tastes in their choices. When you visit an e-commerce site and look for a specific dress, you start seeing several other dresses which are similar. Or, when you watch a video on YouTube, it starts recommending several other videos which are similar. Similarly, when you read a specific news item, the website starts showing you information related to that same category of items. When you search for a book on amazon.com, it also shows other books in the same genre or by the same author.

These suggestions are made by the website, based on what it observed you as taking interest in. It then looks for other users who had shown interest in the same thing, and so, those other users are assumed to have a similar choice or interest as you. Then, it will start making suggestions based on the other items also in which these same users (who have choices similar to you) had shown an interest.

Intuitively, each of the items are characterized by their features or properties that can be measured or classified. Then a score is given to each of the features by an individual and the aggregate score represents the level of interest the individual may have for the item. Practically, data may not be available for an individual's interest in features of item, an item maybe new and features are not known, partial information may exist for other users but not for user under consideration, etc.

While the concepts explained in this chapter work for all of the above situations for recommendation, for ease in explanation, assume that you are building the recommender system for an e-commerce site that sells clothes.

G. Rebala et al., *An Introduction to Machine Learning*,
https://doi.org/10.1007/978-3-030-15729-6_14

14.1 Features Known

Assume that all clothes sellers on your site already provide important features for each of their merchandise. These features could be anything important that a buyer may consider for his/her purchase decisions. Some of the examples could be size, color, manufacturer/brand, type of dress (ethnic wear, formal wear, party wear), price range, and whatever more is of interest.

Let $x_1, x_2, \ldots$ represent each of the features for the dress, based on which a user may make a purchase decision. Assume there are n such features. And, as you had seen in Chap. 3, add in an extra variable x_0, such that its value is always considered 1.

Now, also consider a specific user U, for whom you want to make a recommendation – based on your understanding of his/her preferences. This specific user would not have looked at all the merchandise that your website carries. However, she would have looked at several items, and based on that, you have an idea of which of the merchandise is liked and how much – by this specific user. Assume you can give a score from 0 to 10 for this user's choice for each merchandise that she has looked at.

How you assign a score is not really covered in this chapter. It could be a combination of a user explicitly giving a rating, or the user making the actual purchase, or the amount of time the user spent viewing this specific article.

So, without going into the details of the mechanisms of how the score is obtained, you now have a score for each of the items that U considered. Assume you have such scores for m items.

14.1.1 User's Affinity Toward Each Feature

Assume $\theta_0, \theta_1, \theta_{2, \ldots,}$ etc. are the various coefficients that represent how much value does this user attach to each of features, $x_0, x_1, x_2, \ldots,$ respectively. So, you now have a user U, for whom you know her preference score for m items, and each of these m items have features represented by $x_0, x_1, x_2, \ldots$. You now need to determine $\theta_0, \theta_1, \theta_{2, \ldots,}$ such that for each of these m items, U's preference score is very close to $\sum_{j=0}^{n} (\theta_j * x_j)$.

This is exactly the problem that you had understood as Linear Regression in Chap. 3. Add in the concept of regularization learnt in Sect. 4.4. So, you now have to find the values of $\theta_0, \theta_1, \theta_{2, \ldots,}$ etc. for a specific user U, such that the cost function given by Eq. 4.7 (being reproduced) is minimal:

$$J(\theta) = \frac{1}{2m}\left(\sum_{i=1}^{m}\left(h(x^i) - y^i\right)^2 + \lambda\sum_{j=1}^{n}\theta_j^2\right) \tag{4.7}$$

where:

m: the number of dresses for which you have the preference score for the specific user
n: number of features characterizing a dress
x^i: values of features for the ith dress
$h(x^i)$: predicted value for the preference score for this user for the ith dress, based on $\sum_{j=0}^{n} (\theta_j * x_j)$
y^i: actual preference score for the ith dress – for this specific user
λ: regularization parameter
θ_j: parameter values that you want to determine, in order to minimize the cost function

Since the number of items (m) is fixed, Eq. 4.7 can be further simplified as given in Eq. 14.1:

$$J(\theta) = \frac{1}{2}\left(\sum_{i=1}^{m} \left(h(x^i) - y^i\right)^2 + \lambda \sum_{j=1}^{n} \theta_j^2\right) \tag{14.1}$$

You can now use the Gradient Descent algorithms (Eq. 4.9) to solve for θ_j. By solving for θ_j, you have determined the effective contribution of various features into U's preference. Now that the recommender system knows how much value U attaches to each feature, for any given dress (and the values for its features), it can predict U's preference score and should now present U with only those dresses that have high preference score, and thus are closer to her preferences (expressed in the past).

If you are a huge online store, which carries different kinds of items, e.g., books as well as dresses, the features of interest could be totally different for books, compared to dresses. In such cases, think of the features as the superset of all features considered across all varieties of merchandise being carried.

14.2 User's Preferences Known

Now, assume that when users register on your website, they have to provide their preferences. These preferences could be anything important that a product can easily characterize. Some of the examples could be size, color, manufacturer/brand, type of dress (ethnic wear, formal wear, party wear), price range, and whatever more is of interest.

Let θ_1, $\theta_{2, \ldots}$ represent the preferences of a specific user for various characteristics. For n features, the user will have values θ_1, $\theta_{2, \ldots,}\theta_n$. And, as you had seen in Chap. 3, put in an extra variable θ_0.

Now, also consider a specific dress D, for which you want to decide whether or not you want to recommend this to a user, based on your understanding of this dress's features and the user's choices expressed during registration. Not all visitors to your website would have looked at this specific dress. However, several users would have looked at this dress, and based on that, you have an idea as to how much did each of these users like this dress. Assume you can give a score from 0 to 10 for each (those who have looked at) user's choice for this dress.

So, you now have a score from various users for this dress. Assume you have such scores from *k* users and from their registration details you know their preferences for each of the *n* features.

14.2.1 Characterizing Features

Assume $x_1, x_2, \ldots$, etc. are the various features that characterize each dress, and these are the features for which the users' choices θ_1, $\theta_{2,}$ ... are known (from their registration details). So, you now have a dress *D*, for which you know the preference score from *k* users, and each of these *k* users have their choices represented by θ_1, $\theta_{2,}$ .... You now need to determine $x_1, x_2, \ldots$ such that for each of these *k* users, their preference score toward *D* is very close to $\sum_{j=0}^{n} \left(\theta_j * x_j\right)$. Note that the expression includes θ_0 and x_0 also. For *0*th term, x_0 is known (conventionally always 1), while for all other terms, θ values are known for each user.

This is once again very similar to the problem of Linear Regression but with a slight twist. You now have to find the values of $x_1, x_{2, \ldots,}$ etc. for a specific dress *D*, such that the cost function given by Eq. 14.2 is minimal:

$$J(x) = \frac{1}{2}\left(\sum_{i=1}^{k} \left(h(\theta^i) - y^i\right)^2 + \lambda \sum_{j=1}^{n} x_j^2\right) \tag{14.2}$$

where:

k: the number of users for which you have the preference score for the specific dress
θ^i: parameter values corresponding to various features for *i*th user – possibly from registration details
$h(\theta^i)$: predicted value for the preference score for the *i*th user for this dress, based on $\sum_{j=0}^{n} \left(\theta_j * x_j\right)$
y^i: preference score for the *i*th user – for this specific dress
λ: regularization parameter
x_j: *j*th feature for the current dress

You can now use the Gradient Descent algorithms to solve for x_j. By solving for x_j, you have determined the values of various features for the specific dress *D*. Now that the recommender system knows the features of the dress *D*; for any given user, it can predict this user's preference score for this dress and should now present this dress to only those users, who would have a high value for the predicted preference score given by $\sum_{j=0}^{n} \left(\theta_j * x_j\right)$.

14.3 Features and User Preferences Both Unknown

Section 14.1 provided a mechanism for you to make recommendations when merchandise characteristics were known. Actually, instead of going through all this trouble, you could simply use *K-nearest neighbors* concept, explained in Chap. 6. If a user liked a specific merchandise, identify other merchandise which are its nearest neighbors, and recommend/display these merchandise to the specific user.

Section 14.2 provided a mechanism for you to make recommendations when user preferences were known. In this case also, you could simply use *K*-nearest neighbors concept. If a specific dress was liked by a user, identify other users which are nearest neighbors to this user (in terms of their choices), and recommend/display this dress to these nearest neighboring users.

The concepts explained in the previous two sections provide the building blocks for situations, where neither the user preferences are known well nor are the merchandise features known.

14.3.1 Collaborative Filtering

The basic concept in collaborative filtering is to make use of collaborative information across all users and across merchandise and thus determine information about everything (the merchandise as well as the users).

14.3.1.1 Basic Assumptions

Each of the users have looked at least a few items of the merchandise, and you have the preference score for that user for those specific merchandise.

Similarly, each merchandise item has been looked at, by at least a few of the users, and you have the preference score for that merchandise by these users.

14.3.1.2 Parameters Under Consideration

Consider features $x_0, x_1, x_2, \ldots$, etc. which define the merchandise. So, these are the properties of the merchandise.

Consider parameters $\theta_0, \theta_1, \theta_2, \ldots$ corresponding to the above features. These are properties of the user's affinity toward these parameters.

The combination $\sum_{j=0}^{n} \left(\theta_j * x_j\right)$ represents the preference score for a user (characterized by θ) toward merchandise (characterized by x).

14.3.1.3 Initialize

Initialize each of these features and the parameters to some small random value. Only x_0 is known and is kept at 1.

14.3.1.4 Iterate

While the solution is not converged {

For each user {
Refine the parameters θ_j, based on the current values of merchandise characteristics – considering all the merchandise for which the preference score is known for this user. This can be done using equation 14.1
} // End For

For each merchandise {
Refine the parameters x_j, based on the current values of user choices– considering all the users for which the preference score is known for this merchandise. This can be done using equation 14.2
} End For
} End While

14.3.1.5 Cost Function

The two cost functions, expressed by Eqs. 14.1 and 14.2 can be represented by a single combined cost function. Consider:

N_u: number of users.
N_m: number of merchandise carried by the website.
n: number of features.
$r(i, j)$: a value of 1 denotes that you have the preference score corresponding to user j for the merchandise i, and a value of 0 denotes that such a score does not exist for this user-merchandise combination.
$y(i, j)$: the preference score for user j toward merchandise i. This score has meaning only if $r(i, j)$ is 1.

The combined cost function is given by Eq. 14.3:

$$J(\theta,x) = \frac{1}{2}\left(\sum_{i,j}\left(y(\text{predicted}) - y(i,j)\right)^2 + \lambda \sum_{j=1}^{\text{Nu}} \sum_{k=1}^{n} \theta_k^{j2} + \sum_{i=1}^{\text{Nm}} \sum_{k=1}^{n} x_k^{i2}\right) \tag{14.3}$$

where:

Predicted value for the preference score for user j for the merchandise i is given by $\sum_{k=0}^{n}\left(\theta_k^j * x_k^i\right)$

λ: regularization parameter

θ_k^j: parameter corresponding to kth feature for jth user

x_k^i: kth feature for ith merchandise

A piecewise intuitive understanding of Eq. 14.3 is:

- The first summation expression is evaluated only for those combinations which have $r(i,j)$ value as 1. The expression is the square of the error between predicted value and the actual known preference score and then summed over all combinations of user-merchandise for which the preference score is known. So, this is trying to minimize the error between predicted score and the known score.
- The second summation expression is adding all parameters squared (except θ_0) for all users. Minimizing this expression is part of regularization for the user's preferences.
- The third summation is adding all features squared (except x_0) for all merchandise items and is once again part of regularization.

14.3.1.6 Gradient Descent

Now that you have the combined cost function, you can use Gradient Descent mechanisms to minimize the cost function. Gradient Descent algorithm requires the current values to be adjusted based on the slope (or partial derivative). This adjustment during each iteration is given by Eqs. 14.4 and 14.5:

$$x_k^i = x_k^i - \alpha\left(\sum\nolimits_j (y(\text{predicted}) - y(i,j)) \cdot \theta_k^j + \lambda x_k^i\right) \tag{14.4}$$

$$\theta_k^j = \theta_k^j - \alpha\left(\sum\nolimits_i (y(\text{predicted}) - y(i,j)) \cdot x_k^i + \lambda \theta_k^j\right) \tag{14.5}$$

You should be able to see that these two equations are very similar to Eq. 4.8; and hence, detailed explanation is not repeated here.

The beauty of this mechanism is the features extracted are all derived by the algorithm and you don't even need to worry about the actual physical significance of the features extracted!!!

14.3.2 Predicting and Recommending

Now, you have been able to identify features as well as user's preferences by making use of cumulative learning across all users and all merchandise, based on whatever information (each user providing the preference score for a few of the merchandise).

Now, predicting a specific user's preference score for a specific merchandise is very simple. $\sum_{k=0}^{n}\left(\theta_k^j * x_k^i\right)$ gives the preference score for a user j for merchandise i.

You can also construct a matrix of N_m rows and N_u columns. An element (i, j) of this matrix represents the user j's preference score toward merchandise i. This element value is based on either the preference score known already (and thus used for extraction of features and parameters) or based on predictions (from the parameters and features extracted).

You have already obtained the features for various merchandise items. Instead of worrying about obtaining the preference score for a user toward different merchandise, you can use the *K-nearest neighbors* concept. If a user is showing interest in a specific item, you find other merchandise items which are considered nearest to it in terms of the features and show/recommend these other items also. The advantage of using *K-nearest neighbors* for making recommendations (over, prediction-based recommendation) is that you can modify your recommendation based on the instantaneous interests that the user is showing. This should explain why, when you look for a flight ticket for a specific journey, the sites start showing you other options between the same set of cities. And, on another day, when you look for a ticket for another journey, the information shown to you is for this new journey.

Even though you want to use *K-nearest neighbors* for making the recommendation, you would still need to make use of collaborative filtering technique in order to extract the features.

14.4 New User

Consider an absolutely new user. There is no prior history to know of her preferences or the θ parameters for her, which could indicate her affinity toward various features. What should you recommend to her?

A simple solution is to recommend the merchandise item which has the highest average of preference scores. Consider the matrix shown in Fig 14.1, showing users' preference score for merchandise. As mentioned in Sect. 14.3.2, an element (i, j) of this matrix represents the user j's preference score toward merchandise i. The dashed entry in this matrix denotes that the preference score for this cell (i.e., this user-merchandise combination) is not available. It can be obtained through prediction but is not directly available.

The last column represents this new user, and hence, the entries are all dashes – representing preferences not known. The matrix represents six users (including this new user) and four merchandise items.

Fig. 14.1 Matrix showing preference scores for various users and merchandise items

10	10	0	5	3	-
8	5	-	4	7	-
4	10	7	-	-	-
5	9	8	4	3	-

Table 14.1 Average score for various merchandise

Merchandise seq number	Average score	Calculation
1	5.6	(10+10+0+5+3)/5
2	6	(8+5+4+7)/4
3	7	(4+10+7)/3
4	5.8	(5+9+8+4+3)/5

For calculating the average score for each merchandise, ignore the dashed entries. In that sense, the dash is distinct from 0 – in the sense that 0 score would still be considered (and thus would reduce the average). 0 represents lack of interest, while dash denotes interest not known.

The average score for various merchandise items is given in Table 14.1. As you can see, merchandise item 3 has the highest average rating, and till you learn more about this new user, the recommender system should suggest this merchandise item to this new user.

14.4.1 Shortcomings of the Current Algorithm

For this new user, you want to characterize her affinity toward various features, i.e., you want to find the values of θ corresponding to this new user.

And, you can do it using cost function expressed in Eq. 14.3 (being reproduced):

$$J(\theta) = \frac{1}{2}\left(\sum\nolimits_{i,j}\left(y(\text{predicted}) - y(i,j)\right)^2 + \lambda \sum\nolimits_{j=1}^{\text{Nu}} \sum\nolimits_{k=1}^{n} \theta_k^{j2} + \sum\nolimits_{i=1}^{\text{Nm}} \sum\nolimits_{k=1}^{n} x_k^{i2}\right) \tag{14.3}$$

Consider each of the summation terms. The first term is to be evaluated only for combinations for which preference score is known (or exists). Since this is a new user, the algorithm does not know her preference score for any of the merchandise, and hence, the first term straight away becomes 0. The next two terms being independent and squared need to be minimized independently – in order to get the minimum for the whole cost function. Thus, the solved value for $\theta^{\text{new user}}$ will be evaluated as 0 – for each feature parameter.

With this value of $\theta^{\text{new user}}$, if you now determine her preference score for each merchandise, the result will be the same predicted score for each merchandise, and that score will be 0. So, effectively, the algorithm is not able to suggest anything for the new user. This problem can be solved using the mean normalization technique explained next.

Fig. 14.2 Matrix 14.1 after mean normalization

4.4	4.4	-5.6	-0.6	-2.6	-
2	-1	-	-2	1	-
-3	3	0	-	-	-
-0.8	3.2	2.2	-1.8	-2.8	-

14.4.2 Mean Normalization

For each row, subtract the mean of that row from the cell entries of this row. As before, ignore the dashes. The matrix shown in Fig. 14.1 is now modified to a new matrix, shown in Fig. 14.2. And, now use this matrix for the extraction of features (for each merchandise) and the parameters (for each user).

And, for making a prediction, for a user j for merchandise i, you still use $\sum_{k=0}^{n}\left(\theta_k^j * x_k^i\right)$, but add back the mean for that merchandise that you had subtracted.

For a new user, with this modified matrix, the parameters would still evaluate to 0. And, the predicted score for each merchandise would still compute to 0. However, the moment you add back the subtracted mean, the new score would be the same as given in column 2 (average score) of Table 14.1. Thus, the modified score will suggest that merchandise 3 should be suggested to this new user. Of course, as the algorithm starts learning about her preference scores, the θ values would start becoming more meaningful for this new user also, and recommendations would start getting more personalized.

14.5 Tracking Changes in Preferences

Recommender systems as explained in this chapter are usually needed for online platforms. Assuming a reasonably popular website, you can expect a constant stream of new data. Recollect that in Sect. 4.3, you had learnt about variations on Gradient Descent. Specifically, Sect. 4.3.2 talked about Stochastic Gradient Descent, which helps determine parameters based on per data, rather than looking at the whole dataset.

For an online platform, since there is a constant stream of data, you can use the Stochastic Gradient Descent method, where the parameter values are continuously refined based on the new data, and once a data has been used to refine the parameter values, you can just discard this data.

Say, an existing user gives a preference score to a merchandise. Use this new data to refine the current value of her θ, and discard the data. Similarly, the previous data that were used to arrive at the current value of θ are already discarded. With the refinement that you did to the θ, you no longer know how this new value of θ matches with the prior choices for the same user – since the data corresponding to the prior choices have not been stored.

And, that information (on correlation with prior choices) are not even relevant. In fact, the algorithm now is able to adapt to the changing choices for a user. And, as the user's choice will change, your learned value of θ for that user will also follow the change in user's preferences and choices.

Thus, as a user's lifestyle or choice drift, so do the recommendations.

Chapter 15
Convolution

Convolution is a very important concept in the world of machine learning. In many of the previous chapters, you have read about various algorithms, and you have seen how they work on numbers. Convolution is a technique which automates extraction and synthesis of significant features needed to identify the target classes, useful for machine learning applications. Fundamentally, convolution is feature engineering guided by the ground truth and cost function. Thus, convolution is used for some of the coolest applications of machine learning, such as image recognition, handwriting reading, interpreting street signs, etc. As you can well imagine, one of the most famous applications of machine learning – ADAS (autonomous driver assistance system) – depends on convolution as a component of the whole system to identify objects and to interpret signs!!

In this chapter, you will understand the concept of convolution and will see how to apply this concept to machine learning applications.

15.1 Convolution Explained

Mathematically, convolution can be very complex. For the purpose of this chapter, though, convolution is a technique to mathematically measure the degree of overlap between two figures. Consider the case of convolution – for images as input.

Before you understand the intuition behind convolution, have a look at slightly more formal explanation. An image of 100 × 100 pixels is like a dataset or table of 100 × 100 = 10,000 attributes. The values in each of the 10,000 pixel positions can be a number between 0 and 255 (assuming, 8-bit representation). Depending on the application, some of the pixels may not have any relevant information for the given ground truth, e.g., the ground truth may consist of the task to recognize faces within a 10 × 10 pixels at the center of the 100 × 100 pixels images with a fixed background in all images. If the ground truth requires identifying 100 different objects, then it is almost impossible for human effort to achieve the feature engineering. A

G. Rebala et al., *An Introduction to Machine Learning*,
https://doi.org/10.1007/978-3-030-15729-6_15

convolution layer consists of various set of filters that are tuned by the training algorithm to select a set of attributes relevant to recognize the objects provided as ground truth.

Mathematically, convolution consists of two basic operations:

- Computation of dot product of the filter and the overlapping portion of the image
- Detection of highest overlap from the dot products computed from each overlap of the filter and the input image portion

A convolution filter of size 5 × 5 pixels is the same as 25 attributes relevant to detect parts of object(s) from the ground truth, but the part(s) are algorithmically decided by the training algorithm and that is the most useful advantage of convolution.

For a more intuitive understanding, consider the rectangular shape shown in Fig. 15.1(a), and imagine that the horizontal axis represents time or space (whatever you want). Consider a similar rectangular shape shown in Fig. 15.1(b), except that this shape is at a different position on the horizontal axis. Let this second rectangle slide from the extreme left toward the extreme right. And, the area of overlap is plotted on Fig. 15.1(c).

Till the leading edge (*D*) of the sliding rectangle reaches the trailing edge (*A*) of the first rectangle, the overlap between the two figures is 0. Once *D* reaches *A*, the overlapping area starts increasing. And, once *D* reaches the edge *B*, there is complete overlap. A continued movement of the sliding rectangle now starts reducing the area of overlap. And when the trailing edge (*C*) of the sliding rectangle crosses the leading edge (*B*) of the first rectangle, there is no overlap, and the area of overlap goes back to 0.

By just looking at the Fig. 15.1(c), you can now tell for the sliding figure (rectangle), where the original graph has a corresponding rectangle! So, effectively, by looking at the result of convolution, you can tell the location of a rectangle in the original graph. Or, whether there was a rectangle at all or not in the original graph.

In the above example, you saw a case where the sliding figure exactly matched the figure in the graph. Sometimes, there might not be an exact match. Figure 15.2 shows a case where the match is not exact.

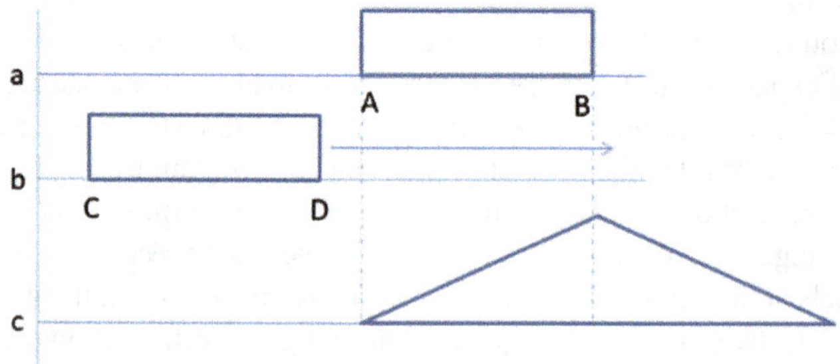

Fig. 15.1 Sliding figure and measuring overlap

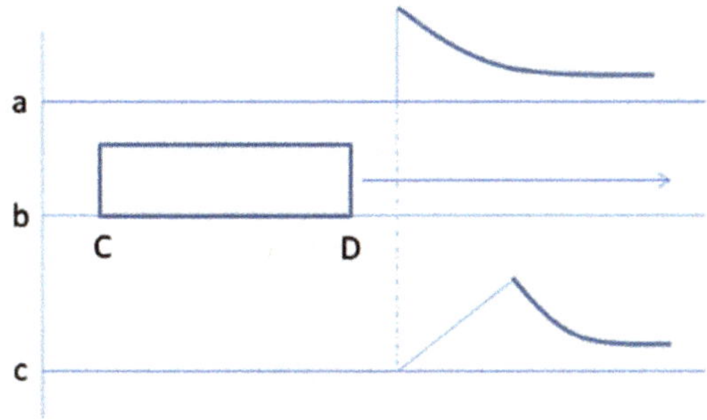

Fig. 15.2 Overlap for nonidentical shapes

So, now, even if the exact shape does not exist, you can determine the extent of likeness of the sliding figure with various locations in the first graph.

The convolution output shown in Figs. 15.1(c) and 15.2(c) is drawn with reference to the leading edge of the sliding figure – for ease in explanation. In reality, these are drawn with the center of the sliding figure. So, the actual output curve would be shifted toward the left (by a distance equal to half the length of the sliding figure). The ref shape which is being searched for in the original image represents the *filter*. Referring to Fig. 15.2, the rectangular shape is the *filter*.

15.2 Object Identification Example

Assume that you are given a picture and you have to identify whether or not there is a small airplane in the big picture.

15.2.1 Exact Shape Known

To start with, assume that you already know the exact shape of the airplane that you are looking for. You will take a small image of the airplane (similar to what you are looking for), and compare this small image with the top left corner, and measure the overlap. Now, you keep sliding this small image horizontally across the original picture, and keep noting down the degree of similarity. The degree of similarity is the dot product between the small image of the airplane and the overlapping portion of the input image. Once you have scrolled across horizontally, you will come again to extreme left, and do the same horizontal scrolling – but this time, slightly below the topmost line, and continue for the whole picture. In this process, you have effectively scanned the whole picture, looking at small pieces at a time and measuring the degree of overlap at each position. If you find a certain position, which has a high degree of overlap (beyond a certain threshold), you know that this position has an airplane similar to the one that you were looking for. Figure 15.3 shows this concept of scanning the entire picture to search for the specific image. Here, you are using the concept of convolution to detect a specific image (or feature) in a larger image.

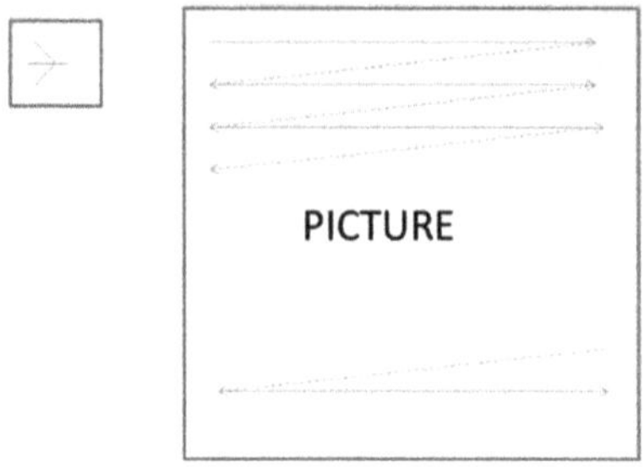

Fig. 15.3 Scanning a picture for an airplane's image

15.2.2 Exact Shape Not Known

In most real-life applications, you want to detect an object (say: an airplane), without knowing the exact specific shape/model of an airplane. In such cases, instead of searching for the image of an airplane, you can search for some specific features/characteristics of an airplane and their relative locations to each other. For example, there should be wings, a pointed nose, a high tail, typically lots of windows (if passenger plane), and often wheels (if on or near the ground). The pointed nose and the high tail should be on the extreme ends of the figure.

So, now, instead of convolving a specific airplane's image, you will convolve for some of these characteristics (wings, pointed nose, raised tail, etc.). If you find each of these shapes and their relative locations (pointed nose and raised tail on two different sides of the wings), and within the same shape, you can say with reasonable certainty that the specific picture has an airplane.

15.2.3 Breaking Down Further

In real-life applications, you don't even have to worry about the specific characteristics of the airplane; you will use some basic shapes.

After convolution of each shape with the original image, you now know which shape lies in which portion of the image. Thus, after the process of convolution, what was originally an image is now transformed into information about which shapes lie in which portion of the image.

If you know which shapes occur in which combinations for airplanes, and if you detect the same combination of shapes in your image, you can say that there is an airplane in the image.

15.2.4 Unanswered Questions

At this point, the basic theory of detecting/identifying an object has been explained; however, there are still a few questions remaining. The important questions that you need to answer before you can actually do an object detection/identification are:

- Exactly how do you convolve a shape/image with another (larger complete) image?
- How do you know which combinations of which shapes will constitute an airplane (or any object that you are interested in)?

15.3 Image Convolution

Before you go into understanding the process of image convolution, let us understand why you need such a complicated mechanism for detecting an object. After all, you can detect airplanes without even exercising your brain. Even a toddler can. So, why all this effort for a computer?

The reason is what you see as a complete image, computer does not. It sees an image as a combination of intensity for each pixel. And, this is typically three dimensional. The two dimensions are standard dimensions to represent a two-dimensional plane and an additional dimension for each color (red, green, and blue). Figure 15.4(a) shows an image (image courtesy: www.disastermgmt.org), and Fig. 15.4(b) shows the pixel representation (grayscale) for a small segment (shown in the box) of the picture.

While this representation of an image as a bunch of numbers creates a challenge for computer to identify the whole image, this also provides a way for you to perform the convolution.

You want to search for a given shape (called: *filter*) in a given larger image. Each of these is represented by a set of numbers – denoting intensity values for the pixels. Since the shape (*filter*) and the image are both two dimensional (for each color), the pixel intensity can be written in the form of a matrix.

Equation 15.1 provides a measure for the degree of match at any position:

$$\text{Overlap} = \sum F_i * I_i \tag{15.1}$$

where:

F represents the filter matrix
I represents the corresponding portion of the image matrix
i is varied over all the elements of the matrices

The *overlap* of 15.1 is a computation of degree of similarity or the dot product. A dot product of two vectors is the degree of overlap between two vectors. In Eq. 15.1,

a

b

40	36	39	39	34	33	37	37	37	38	45	45	44	40	40	40	43	37	35	34
36	35	34	33	33	37	40	38	38	50	64	73	73	61	48	41	39	35	34	34
34	36	32	32	37	39	38	41	49	90	107	120	121	107	79	54	37	36	35	35
32	38	33	36	44	44	40	56	81	134	141	144	143	131	102	63	35	37	36	37
34	38	35	42	54	52	50	78	117	156	146	135	132	126	103	64	31	40	38	40
41	54	31	58	72	52	59	107	167	150	143	153	115	116	124	62	39	42	42	43
40	48	37	57	61	57	79	129	187	181	194	182	161	141	161	82	44	55	37	32
58	54	79	108	104	105	125	148	185	219	219	201	193	189	184	127	88	66	64	90
62	78	137	182	164	157	165	171	193	215	177	168	161	203	152	150	130	111	123	170
84	101	170	213	182	162	174	188	216	206	162	150	142	187	147	161	156	167	172	200
72	122	179	222	192	165	172	186	208	200	192	168	159	161	180	168	167	175	191	211
118	150	180	225	214	187	188	187	188	172	178	166	148	137	184	159	169	177	201	204
164	167	175	217	223	207	213	208	200	162	150	164	135	142	172	163	191	210	209	169
160	148	126	150	196	212	214	214	210	169	171	194	190	174	175	176	184	217	179	133
176	113	79	92	150	187	188	189	205	188	162	165	166	160	164	166	180	187	159	140
163	90	60	56	103	154	164	159	175	190	151	151	164	171	176	175	190	177	166	152
126	72	69	64	88	148	176	166	166	175	156	169	178	177	176	170	178	183	185	160
97	62	66	68	94	158	192	188	192	162	165	184	167	142	139	140	151	192	189	173
78	71	55	68	119	174	188	190	207	163	170	184	152	119	129	145	164	213	194	205
83	81	61	89	154	195	197	188	189	172	169	182	165	148	161	177	196	222	205	223

Fig. 15.4 (**a**) A sample image. (**b**) Pixel intensity values for a portion of the image

it is a degree of overlap between two sets of vectors. To better understand the equation, consider each element of the filter matrix, and for each element multiply it with the corresponding position's element of the image. So, now, for a *n-by-n* filter, you have *n-by-n* product elements. Add all these elements, and this represents a degree of match of the specific filter with that position of the image. Note that this operation is different from matrix multiplication.

a

a	b	c
d	e	f
g	h	i

b

m	n	o
p	q	r
s	t	u

c

am	bn	co
dp	eq	fr
gs	ht	iu

Fig. 15.5 (**a**) Pixel representation for a 3-by-3 filter. (**b**) Pixel representation for a 3-by-3 portion of an image. (**c**) Result of pixel-by-pixel multiplication

Figure 15.5(a) gives a sample representation of pixels for a *3-by-3* filter. Here, *3-by-3* indicates a 3 pixel by 3 pixel shape. Fig. 15.5(b) gives a sample pixel representation for a *3-by-3* (i.e., same size) portion of a much larger image, and Fig. 15.5(c) represents the multiplied values for the pixels from the two matrices. The notation *am* in this figure represents algebraic expression meaning product of values *a* and *m*. Finally, all the values *am* through *iu* of Fig. 15.5(c) are added (see Eq. 15.2) to get one single value representing the overlap. While you say that this output value represents the overlap at this location, it is actually considering *nine* positions of the image. For the purpose of convolution, the position is considered to correspond to the center pixel. Once again, remember that the overlap is not matrix multiplication. It involves pixel-by-pixel multiplication, as given by Eq. 15.2.

$$\text{Overlap} = am + bn + co + dp + eq + fr + gs + ht + iu \tag{15.2}$$

15.4 Preprocessing

If the aim is object detection or classification, many times, you may not even need the whole details of a solid picture. In many cases, just the outline of the objects in the picture will be sufficient.

In such cases, before applying the convolution, you may want to apply edge detection on the given image.

Similarly, if the color of the objects is not of importance, you may convert the color image into a black-and-white or grayscale image.

The advantage of these preprocessing steps is that you are now reducing data volume by a huge factor. By converting the color image into grayscale, instead of storing and processing *three* different intensities (red, green, and blue), you will be working with only a single intensity (grayscale). This directly reduces the problem size by a factor of 3. Further, by retaining only the edges, you will be darkening all other points which were part of a solid object. These dark pixels will be represented by 0. You now have sparse matrix, since all internal elements of solid figures are

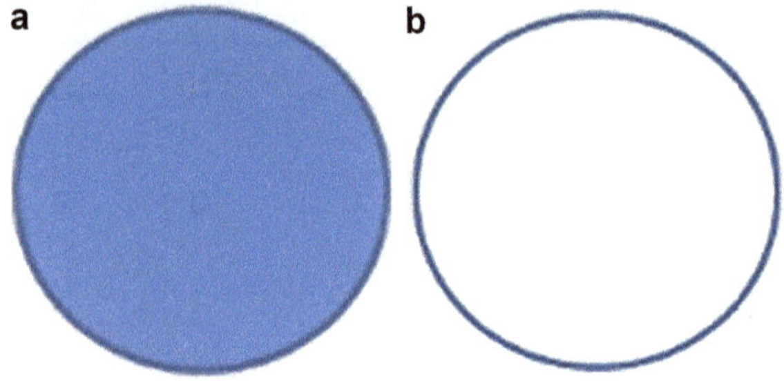

Fig. 15.6 (**a**) A solid circular shape. (**b**) Only edges of the circular shape

now 0. Just for fun, Fig. 15.6(a) and (b) represents a solid figure and only the edge, respectively. You can see that:

- Both are equally good, when it comes to detecting that these are circles.
- Figure 15.6(b) has many more blank pixels and, hence, will be sparse matrix.

Sparse matrices will have following major advantages:

- As you saw in the previous section, there are a lot of element-by-element multiplications. Having many elements at 0 is very helpful for such multiplication.
- Sparse matrices can have their own data structure, which are very efficient. You will see in Sect. 15.8 that the image data is represented in a form, where each pixel is repeated several times. Having a lot of 0s in the original image representation will cause many more 0s in that modified data representation also.

15.5 Post-Processing

You now have for a given filter the degree of overlap of that filter with each location of the image. You can do some post-processing to reduce the data size further. Consider any position, and look at all its surrounding pixels. You can do a *max* of the *overlap* for all these pixels, and use this max value as the *overlap* value at this location. This concept is called *max pooling*. Or, you could do an average and use this average. This is called *average pooling*. Similarly, you could also use *min pooling*. It is also not necessary that you have to look at only the immediate neighboring pixels for pooling; you can decide to look for all pixels within *two* positions away (or even further).

An advantage of such pooling is that these post-processing steps also provide some kind of smoothening of your results. Assuming *max pooling*, a very high value of overlap will bleed into surrounding pixels. This means that, for the shape of interest, you are no longer dealing with the exact location, correct to the pixel. Rather, you have now localized it to the general area, at and around that position. Further, pooling will allow you to make use of something called *stride*, which you will see in the next section.

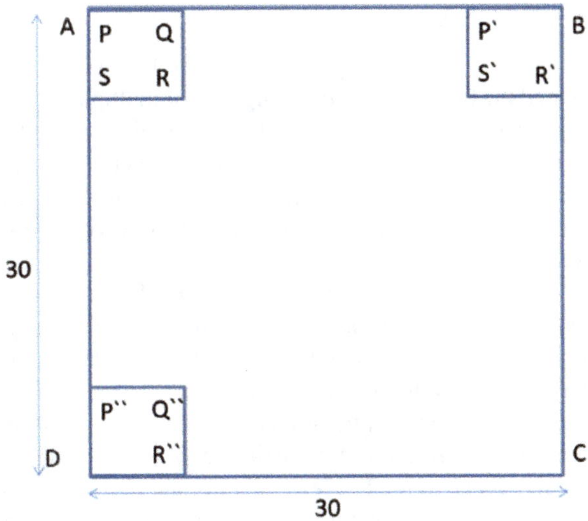

Fig. 15.7 Data reduction due to convolution

The output of convolution is then fed into a neural network, with the whole system being called *convolutional neural network* (or *CNN*). The process of convolution will detect which shape is found at which location, and neural network will tell, given the presence of various shapes at various locations, what is the object involved in totality. In Sect. 15.9, you will further see that convolution itself is a specialized form of neural network.

15.6 Stride

Consider a situation shown in Fig. 15.7. *ABCD* represents an image of size 30 × 30. *PQRS* represents a filter of size 4 × 4. You intend to convolve this filter over the image. You will start with *PQRS* being placed at the extreme top left and will keep sliding horizontally till it reaches the extreme right of the top, represented by *P`BR`S`*. Satisfy yourself that you will get only 27 positions (reduced from 30). And, you will get this reduction while convolving each row. For the last row, the filter will start at location represented by *P``Q``R``D*. Hence, there will be reduction of three rows also.

In the above paragraph, you assumed that the filter is being moved by *one* position each time while moving right, and when going to the next row as it moves down by *one* position. It should be possible to move it by several positions in each go. The number of positions moved during each step is called *stride* length. So, if the stride length is 2, you will only have half the number of output values. Thus, concept of *non-1* stride length reduces data volume by factor of stride length. If you are doing a *max pool*, where a high value will bleed into its neighboring cells, there is no point in first computing each value, and let the max value percolate. You might as well jump in steps.

15.7 CNN

A CNN network is usually composed of multiple layers of convolution and pooling combination and then followed by a neural network. Convolution itself is a special type of neural network, as explained in Sect. 15.9. The network is defined by the number of the filters, the stride lengths, the number (and sequence of) convolution pooling combinations, and the neural network. Figure 15.8 represents one such network.

The input image is represented as *three* films. These *three* films are pixel intensity values for each of the *three* colors (red, green, and blue). Assuming, no conversion to grayscale, you will have these *three* films (or layers), representing an image. After the first convolution step, your output is a *three*-dimensional shape. The face (representing first layer) of this three-dimensional shape is smaller than the size of the original image. The reduction in face size is due to stride length, and a small reduction due to the last few columns and rows not being straddled, as explained in Sect. 15.6. However, this *three*-dimensional shape is much deeper (i.e., has many more layers), compared to the original image having a depth of just *three*. So far, in our discussions, we have been talking about locating a specific shape. However, you will be trying to locate many kinds of shapes. And, each such shape (i.e., filter) will have a layer (or *three* layers for colored filter). So, if you have 50 filters, the depth at first stage will be 50 or 150 depending on whether you are dealing with single color (grayscale) or colored image. Each layer in this three-dimensional shape represents the position-wise overlap of specific filter in the image.

The next stage is pooling. Here, the number of layers remains unchanged. However, multiple data points are pooled into a single data point, thus further reducing the size of the face.

You can do several such layers of convolution and pooling. And, the final *3D* shape will have a smaller face but will have more depth. Each element in this *3D* shape will now feed into the input layer of the neural network.

For training your network, you take a large number of labelled images. To start with, you will have random shapes for the filters. Convolve each of these filters over

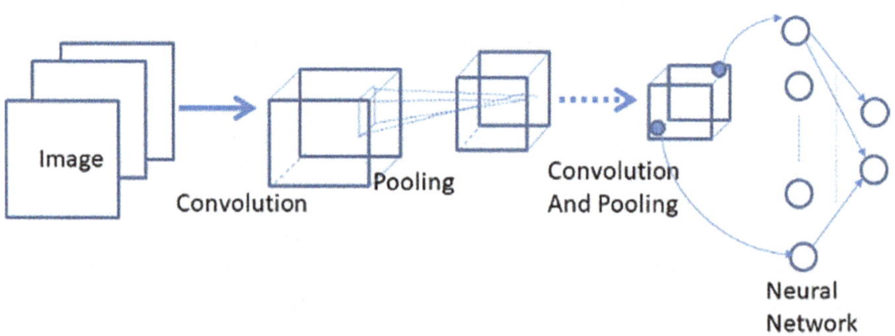

Fig. 15.8 A convolutional neural network (CNN)

your image. And, do pooling. Feed these into your neural network, and keep adjusting the weights till the network is trained. Chapter 9 explained on adjusting weights through back propagation mechanism. Section 15.9 explains how the same concept can be extended to refine the filters also, which were random to start with. Obviously, you need to train your network with a huge number of labelled data, in order to arrive at the correct set of filters and the weights for the neural network.

One of the most difficult questions that comes up is, while there are algorithms for adjusting the weight, how do you come up with the right network, in the sense, the number of convolutional and pooling layers, whether to do *max pooling*, *min pooling,* etc.? Unfortunately, there is no straightforward algorithmic based answer to these queries. As you start playing with various kinds of networks, gradually, your intuition will help you guide on additional refinements that you should try. One good starting point is to implement and experiment with already defined/published networks (e.g., *AlexNet*).

15.8 Matrix Operation

Since there are too many convolution operations, even a slight improvement in your algorithm will have a huge impact on the overall performance of your algorithm. While implementing convolution, you can exploit matrix multiplication for a faster result. This will allow you to obtain many results in one multiplication operation.

Consider the filter and the image shown in Fig. 15.5(a) and (b) once again. Represent a, b, c, ..., i as a row of the first matrix. And, represent m, n, o, ..., u as a column of the second matrix. Multiplying these two matrices gives the same result as Eq. 15.2, i.e., convolution output for the filter being at this position. Other filter values can be put in as additional rows of the first matrix. Thus, the corresponding column of the resultant product matrix now provides the convolution output for all the filters for this position of the image matrix.

So, your 50 filters are represented by 50 rows of a filter matrix. Multiplying this filter matrix with the column (corresponding to one location on image) will give back 50 values in a column. Visually, considering Fig. 15.8, each of the values in this column represents one value on each of the 50 layers of the resulting three-dimensional shape.

To start with, point u of the image is multiplied with point i of the filter. As the filter moves to the right, the same point u is multiplied with point h of the filter and, then, with point g. And when the filter is moved vertically, u will be multiplied with f and then again with e and d and so on. So, you can now create additional columns in the second (image) matrix, where each element is shifted slightly to represent multiplication by the filter getting shifted. Thus, the image data is represented by a modified matrix, where each of the point occurs multiple times (e.g., 9 times for a 3×3 filter), and data is arranged in a manner, such that the product of two matrices, the filter matrix and the modified image matrix, is the desired convolution output.

Effectively, instead of moving the filter (to the right and down), you are moving the image (towards left and up).

Arranging the image data into such a matrix can be extra useful if your underlying hardware supports high degree of parallel processing, such as GPU or FPGAs.

15.9 Refining the Filters

By now, the only part left is how to come to the right set of filters. As mentioned in Sect. 15.7, you start with a random choice of filters, and then, you can refine the filters. Random choice here means assigning random values to the pixels in the filters. This section explains how to refine these random values in the filters.

Convolution can be seen as a special type of neural network. Consider the original image as the first layer of the neural network and the convolution output as the next layer. Consider once again the example data shown in Fig. 15.5(a) and (b). The convolution output is $am + bn + co + dp + eq + fr + gs + ht + iu$. Among these, $m, n, o, \ldots, u$ are the input nodes, and $a, b, c, \ldots, i$ are the weights.

So, convolution can effectively be seen as a slightly specialized kind of neural network, where:

- The output (convolution result) is obtained by only a few nodes of the previous layer. For example, for a 3-by-3 filter, for each (convolution output) node, only nine nodes of the previous layer feed into it.
- Each weight will not be unique. For example, weight a will be common across multiple connections. For example, filter value a will be multiplied by image pixel value m for the first node of the second layer. And then, the same filter value (a) will be multiplied by next pixel (n) of the image for the next node of the second layer.

  Figure 15.9 shows how same weights are used for multiple connections.

Like any other neural network, for this convolution-type neural network also, the weights need to be adjusted during the training phase. These adjusted weights thus become the filters to be used for inference.

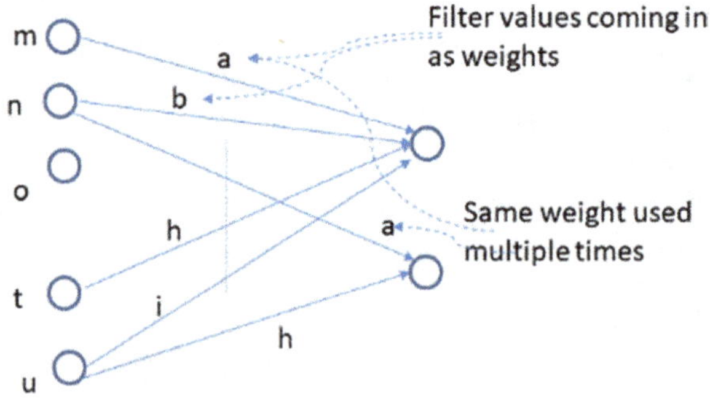

Fig. 15.9 Convolution seen as a neural network

15.10 Pooling as Neural Network

Just like convolution can be implemented as a neural network, pooling can also be implemented as a neural network, except that this is much simpler. Consider pooling for 1 pixel and *average pooling*:

- Once again, only a few elements of a layer will feed into each element of the next layer.
- For average pooling, each of the weights will be constant, e.g., considering values within a single pixel, each element will contribute only 1/9 of its value to the final average value (final value = average of 9 pixel values).
- The node itself only needs to add all incoming net values, and no other nonlinear function has to be applied.

For max pooling, the weights on all the branches will be 1. And, the node will be performing a max function.

15.11 Character Recognition and Road Signs

By now, it should be clear on how to read/recognize characters – including handwritten ones. For the sake of completeness, for typed characters, convolve the filters (each character of interest could be a filter) over the text that you want to read, and areas of high overlap will indicate the presence of that character, at that position. For handwritten characters, it will be better to learn filter shapes, rather than the typeset characters.

Once you know how to recognize characters, you can also process and interpret written road signs such as speed limits, etc. You can also use convolution to detect specific road signs, such as *STOP* signs, without having to actually read the text on those signs.

Sometimes, you may have to do scaling before convolution, e.g., you want to read a number plate from a vehicle's image. Once you have extracted the characters on the number plates, you may have to scale the image so that the height of the characters matches the height of your filter characters – so that convolution would provide a good overlap for matching characters.

15.12 ADAS and Convolution

Convolution (along with neural network) forms the basis of autonomous driver assistance system; the cameras feed into the ADAS system, which detects objects, such as signs written on the road (OK to go over those) versus objects such as cars and pedestrians (avoid coming too close to these), versus signs posted overhead –

which will not come in the way of the vehicle. The ADAS system also interprets road signs, such as speed limits, *School Zone Ahead,* etc., to adjust the speed of the vehicle, based on the signage.

ADAS is one of the most prominent applications which has increased the interest in and awareness of machine learning. The detection of objects and road signs (through convolution learnt in this chapter) is then followed by the actual learning of driving actions and that concept is covered in next two chapters on reinforcement learning.

Chapter 16
Components of Reinforcement Learning

Reinforcement learning (RL) is one of the fundamental areas of machine learning and is widely regarded as a necessary component of artificial general intelligence (AGI). While RL has been researched for a few decades, the advent of deep learning has resulted in the so-called deep reinforcement learning algorithms that utilize deep neural networks and large-scale computing power to significantly improve the capabilities of RL. They have resulted in computer programs that play video games and board games at super human level and even beat handcrafted computer programs, e.g., chess program *AlphaZero* beat the best handcrafted program *Stockfish 8*. Besides video and board games, RL is being applied to enable robots to learn complex tasks in autonomous driving and industrial automation. This chapter covers the fundamental concepts of RL and the taxonomy of RL algorithms. In the next chapter, you will learn some of the key algorithms in reinforcement learning. An understanding of this chapter will provide you with a good feel for RL.

16.1 Key Participants of a Reinforcement Learning System

You need to understand the key components of a reinforcement learning system before delving into the algorithms.

16.1.1 The Agent

The *Agent* is the computer program that is learning and is what you will design. The *Agent* observes, interacts, and modifies its *environment* over time. The *environment* can be modified by the *Agent* or by something else. The Agent perceives the *environment* as a set of observations.

G. Rebala et al., *An Introduction to Machine Learning*,
https://doi.org/10.1007/978-3-030-15729-6_16

Example 1 A chess playing *Agent* must observe the chess board and make chess moves. The *environment* in this context is the chess board configuration (i.e., positions of various chess pieces on the board and whose turn it is) during the game. The *Agent* keeps observing the changing environment. The *Agent* can modify the *environment* by making a chess move. Or, the *environment* can be modified by the opponent, who is also making chess moves. The opponent is usually a human player or another Agent.

Example 2 An *Agent* driving an autonomous car observes the physical surroundings such as roads and traffic while also tracking the car's current location and velocity. Therefore, the *environment* would be the physical surroundings as well as the car's location and velocity. The *Agent* is modifying the *environment* whenever it drives the car to a new location or changes the car's speed. The *environment* can of course change without the *Agent's* input. The *Agent* views the *environment* through a set of sensor readings and images.

Example 3 For a robotic system, the *environment* includes the physical robot itself such as its motors and joints, as well as the physical environment the robot is in. The observations from the *environment* are a set of joint positions, motor speeds, sensor readings, and the images from the physical environment.

An *environment* can also be (created by) another computer program, such as a flight simulator, when the *Agent* is learning to fly a plane.

The *Agent* interacts with the *environment* in a closed loop, as per the following sequence – also shown in Fig. 16.1.

Step 1: *Agent* observes the *environment* (e.g., chess board positions).

Step 2: *Agent* chooses an action from a set of possible actions based on the *environment* and then performs that action (e.g., make a chess move). More commonly we say that *Agent* passes that action to *environment* (to be executed by the *environment*).

Step 3: *Agent* goes in a *wait* state for the *environment* to send new observations (e.g., *Agent* waits for the board configuration to change due to opponent's chess move).

Step 4: *Environment* executes the action it received from the *Agent* and issues new observations to the *Agent* (e.g., the *environment* changes the board configuration due to *Agent's* action; then, when an opponent makes the move, it passes the new board configuration to *Agent*).

Step 5: Go back to step 1.

Since the interactions between *Agent* and *environment* happen sequentially, there needs to be a notion of time to describe the interactions over time. Similar to a wall clock, the system (*Agent* + *environment*) uses a clock (implemented as a counter in computer program) that starts at time 0 and increments by 1 timestep. The system starts executing at time *zero*. Once the *agent* receives an observation, the time is incremented by 1 timestep before it receives the next observation, as you will see in Sect. 16.1.3.

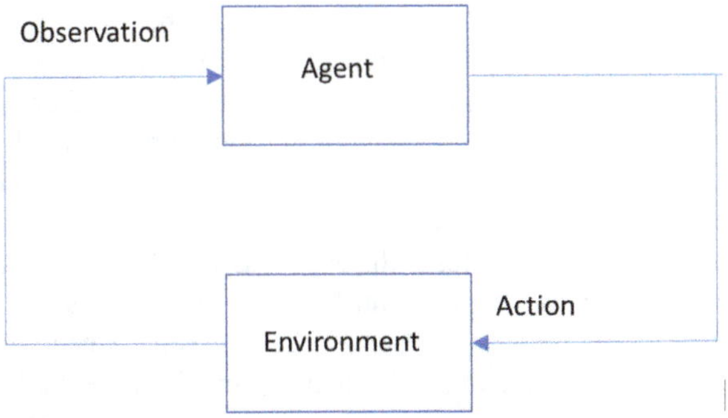

Fig. 16.1 Agent interaction with environment via actions and observations

The actual elapsed time between successive timesteps can vary depending on the problem. For example, in chess play, timestep is the time between two successive chess moves of the *Agent* and that can be seconds or minutes, while, for autonomous driving, the timestep may be 10s to 100s of milliseconds.

Note that we are limiting time to discrete values (0, 1, 2, . . .). *Continuous time environments* where time can take on any real values could be handled by discretizing the time (i.e., have discrete timesteps with small enough time increments).

16.1.1.1 Agent's Objective

The *Agent* always has a well-defined objective. All its actions (toward changing the environment) are geared toward achieving that objective. For example, the objective of a hill climbing robot *Agent* is to reach the hilltop in fewest number of steps without falling.

16.1.1.2 Rewards as Feedback for Agent

The *Agent* usually starts at time 0 with no knowledge of *how* to achieve the objective. There are no training examples for the *Agent* to learn from. Instead the *environment* helps an *Agent* learn by giving occasional feedback based on the Agent's past actions. This feedback is known as *reward* or the *reward signal*. It is a numeric value and usually a real number. A negative value is sometimes known as *punishment*.

The *Agents* are programmed with just one objective: Issue actions that maximize the cumulative rewards obtained from *environment* over time. While *Agents* may have different task-specific objectives (win games, climb mountains, drive cars, etc.), from algorithmic perspective, there is just one objective – maximize cumulative rewards. So you need to design a *reward structure* for the task-specific objective. A *reward structure* defines how the rewards are issued by *environment*. For example, to maximize the number of steps a robot takes (such as distance traveled), give it a positive reward for each step. On the other hand, to minimize the number of steps it takes to perform a task, give it a negative reward for each step. You can also vary the reward amounts in proportion to how good or bad an action is. For example, a robot falling down while walking should result in a large negative reward.

The reward an *Agent* receives is usually delayed, i.e., the reward it receives at current time may be due to its actions going back several timesteps. The *environment* usually does not inform the *Agent* as to which of its actions resulted in rewards. Finding which actions contributed to the rewards is called the *credit assignment problem*. It is an important problem to solve since it helps an *Agent* learn which actions are good and bad for maximizing future rewards. The Agent has to solve the *credit assignment problem* from the history of actions, observations, and rewards.

The rewards can be *sparse*, i.e., the agent may only occasionally receive a reward. For example, instead of receiving a reward at each timestep, it may receive a single reward at the end of the task (positive or negative, depending on whether the task was successful or unsuccessful). In board games such as Chess and Go, the rewards are only given at the end of a full game. Dealing with sparse rewards can be a challenge for *Agents* since it makes the *credit assignment problem* even harder.

16.1.2 The Environment

You already have some feeling of what an *environment* is, based on discussions in Sect. 16.1.1. More formally, it has a mathematical model known as *Markov decision process* (*MDP*). The model is similar to a state machine in computer programming. Before understanding an *MDP*, you need to understand its ingredients – states, the state transition probabilities, rewards, and actions.

An *environment state*, or simply a *state*, is a numerical description of an environment at any given time. The *state* is usually represented by a set of features called the state variables. The values of these variables form a state vector that represents the *environment state*. As an example, consider an *Agent* in a robot, attempting to travel to a target location. Its environment *state* must include at least these features:

- The location (i.e., coordinates) of the robot, represented by a state variable
- The velocity (speed and direction) of the robot, represented by a state variable
- The physical posture of the robot represented by the positions of various joints
- The distance to the target, represented by a state variable

The set of numeric values for these features (state variables) at time t makes up the environment's *state* at that time. The *state* thus characterizes the *environment*. You say that an *environment* is in a particular *state*. When the *state* changes, you say that *environment* has *transitioned* from one *state* to another *state*.

16.1.2.1 Environment State Space

The number of possible *environment* states is the product of the number of possible values for each of the feature variables in the state vector. For example, if you have three features in the state vector and one of them can take 10 possible values, and each of the other two can take 20 possible values, then total number of states is 4000 (=10 * 20 * 20).

If a feature variable takes on *real* values, then the number of possible values for that feature is infinite. In practice, though, the possible values are limited by the resolution at which you can measure quantities of interest or the memory required to store the numbers.

Real-valued state variables are referred as *continuous state variables*. State variables that can only assume a fixed set of values are *discrete state variables*.

The set of all possible states is known as *state space*. The state space for an environment where all states are discrete states is a *finite state space* or *discrete state space*. The state space for an environment with continuous states is a *continuous state space*. In practice you would approximate a continuous state space with a discrete state space by restricting the continuous state variables to take on a fixed number of possible values.

This allows you to solve many reinforcement problems with continuous state space using the methods for discrete state spaces. The process of converting a continuous variable to a discrete variable is known as *discretization*.

16.1.3 Interaction Between Agent and Environment

Figure 16.2 is a refinement over Fig. 16.1. It shows that the *Agent* receives the reward *and* state from environment and outputs an action. The environment receives action from Agent and outputs reward and state for the Agent.

Time is incremented after *environment* receives the action and before it issues the next *state* and *reward*. So, effectively, the system (Agent + environment) produces a time-ordered series of rewards, states, and actions: $r_0\, s_0\, a_0\, r_1\, s_1\, a_1 \ldots r_t\, s_t\, a_t \ldots$. Any

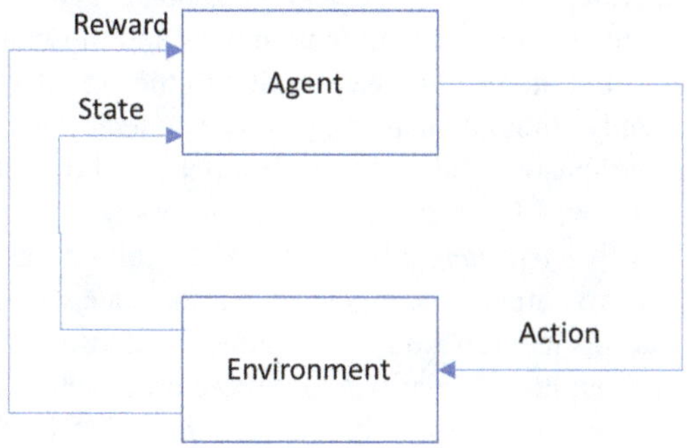

Fig. 16.2 Interaction between agent and environment: refined

such sequence of rewards, states, and actions is known as a *trajectory*. $R(0)$ is the initial reward (usually 0), and $S(0)$ is the initial state of the environment which can be any state.

A *trajectory* can either terminate or go on forever. It can terminate, for example, when an objective is achieved (robot has reached the top of the hill) or when giving up on an objective (robot falls and cannot recover). When terminated, the last state is known as *terminal state*. A full trajectory from initial state to a terminal state is known as an *episode*. For example, a chess game played by an Agent from start to finish is an episode. An *Agent* is said to perform *episodic tasks* when it is involved in an episode.

An example of nonterminating trajectory would be industrial robotic applications that are never turned off. In such situations, Agents (e.g., inside the robots) are said to perform *continuing tasks*.

The trajectory from time 0 to time $t - 1$ is the *history* available to *Agent* at time t. The Agent can use this *history* toward achieving its objective.

16.2 Environment State Transitions and Actions

The environment can change with time through *state transitions*. The state transitions are always caused by an action. However, not all actions may lead to change of state. There are a fixed set of allowable actions from each state (e.g., there are a fixed set of legal moves from a given chess board configuration). The environment's behavior over time can be either deterministic (i.e., predictable) or stochastic (i.e., random).

16.2.1 Deterministic Environment

In a deterministic *environment*, for any given *state*, it is possible to predict the outcome (next state and the reward) due to an action. One example of a deterministic *environment* is the *Agent* playing board games such as Chess and Go where the actions involve making legal moves and those actions change the board configuration in a predictable manner due to the Agent's move. The opposite player subsequently, though, may play in a nondeterministic manner. The rewards are also deterministic. There may be just one reward for the Agent, at the end of the game: +1 for win, 0 for draw, and −1 for loss.

This *environment* behavior can be fully described by a function with two inputs and two outputs: the inputs are current state s_t and action a_t, and the outputs are next state s_{t+1} and reward r_{t+1}. The function doesn't depend on time. The subscripts t and $t + 1$ are just for clarity and represent the values at those times.

You can represent this function in an MDP table, as shown in Table 16.1.

Table 16.1 An example MDP table: deterministic environment

Current state	Action	Next state	Reward
s_1	a_1	s_2	r_2
s_3	a_4	s_1	0
s_4	a_0	s_6	r_5
s_{16}	a_2	s_{16}	0

The last row shows an example where state does not always change after taking an action. A terminal state can be modeled by having all actions for it map back to itself with zero reward.

16.2.2 Stochastic Environment

In a stochastic environment, it is not possible to exactly predict the outcome of an action from any given state because multiple transitions are possible from the same state. Each possible transition has a probability. This probability is called the *state-action transition probability* or simply *state transition probability*. This probability would be an additional column in the *MDP table*, as shown in Table 16.2.

The first two rows demonstrate two possible transitions from s_4, on the same action a_1. The environment will transition to one of s_6 or s_8 with certainty since their probabilities add up to 1, though you can't predict the exact transition among these two states. Suppose there are *N* rows for a particular *state-action* pair (s_i, a_i), then, the sum of probabilities in those *N* rows must add up to 1.

Stated more formally, in a stochastic MDP, every state-action pair maps to a *probability distribution* of next states and rewards instead of a single state and reward.

The *probability distribution* of next state and rewards for a state-action pair (s, a) is simply a list of triples: (s_0, r_0, p_0), (s_1, r_1, p_1), ..., (s_N, r_N, p_N), such that the probability of $s_i.r_i$ pair is probability p_i and the sum of probabilities $p_0 + p_1 + \ldots + p_N$ equals 1.

When environment is in state *s*, and it receives an action *a* from the *Agent*, it looks up the distribution for (s, a) and *samples* this distribution to determine the next state and reward, to be passed on to the *Agent*.

Following the standard convention for probability distributions, let S_{t+1} and R_{t+1} denote the random variables for next state and reward. They take on specific numeric values with certain probabilities, as listed in the triples.

The probability distribution of next states and rewards for a specific state *s* and action *a* in the standard conditional probability notation is expressed as:

$$
\begin{aligned}
\Pr(S_{t+1} = s_0, R_{t+1} = r_0 | S_t = s, A_t = a) = p_0 \\
\Pr(S_{t+1} = s_1, R_{t+1} = r_1 | S_t = s, A_t = a) = p_1 \\
\Pr(S_{t+1} = s_N, R_{t+1} = r_N | S_t = s, A_t = a) = p_N
\end{aligned}
$$

Since this is a probability distribution, $p_0 + p_1 + \ldots + p_N$ should be 1.

Table 16.2 An example MDP table: stochastic environment

Current state	Action	Next state	Reward	Probability
s_4	a_1	s_6	r_2	0.3
s_4	a_1	s_8	0	0.7
s_1	a_0	s_6	r_5	1

The first row is read as follows: Given state s and action a, choose next state s_0 with reward r_0, with probability p_0. Using this convention, the first row of Table 16.2 would be written as $\Pr(S_{t+1} = s_6, R_{t+1} = r_2 | S_t = s_4, A_t = a_1) = 0.3$.

You can describe the whole environment, using either the table style shown in Table 16.2 or using probability distribution shown above.

The pseudo code for environment's behavior at time t is given as:

Read action a(t) from Agent
(Let s(t)= current state)
Obtain probability distribution Pr(S,R | S = s(t), A = a(t)) from the MDP table
Sample the distribution to get actual values for next state sNext and reward rCur
Increment time t
Send next_state sNext and reward rCur to Agent

The stochastic *MDP* can be stationary (probabilities are fixed over time) or nonstationary (probabilities changes over time). Most of our discussion applies to stationary MDPs. However, some of the methods described also apply to nonstationary *MDPs*.

16.2.3 Markov States and MDP

So far you've seen an environment where a state transition at time t only depends on the state and action at time t and doesn't depend on the history of state changes or actions up to time $t - 1$. Such states are known as *Markov states*. An example of a non-Markov state would be a state s5 such that transition from state s_5 to state s_{16} at time t is allowed only if the previous state is s_4.

For reinforcement learning we only consider environments with *Markov states* since they are easier to analyze and apply and also represent the many real-world situations. A classic example is a robot moving to a target location. Its current state is its current position and velocity. Its future movement from here does not depend on its history of movements. All that matters is its current state. Another way to think of this is that current state already incorporates the relevant past history (accumulation of previous positions).

The model of a stochastic environment with Markov states, actions, rewards, and state transition probabilities is known as *Markov decision process* or *MDP*. Before you attempt to apply the reinforcement learning algorithms to an environment, make sure that the environment *can* be modeled using an *MDP*, even if the actual *MDP* model is unknown.

16.3 Agent's Objective

An Agent's objective is to maximize the cumulative reward over time.

In episodic tasks the objective is to maximize the total reward obtained in the episode. Consider an Agent in the middle of an episode, say at time t. It can only maximize the cumulative future reward for the rest of the episode, since it cannot go back in time to change the past rewards.

Let $R(j)$ denote the reward obtained at time j.

Define *return* $G(t)$ as the cumulative reward Agent receives for the rest of the episode, starting at time $t + 1$, as given by Eq. 16.1. By this definition the total reward for an episode is $G(0)$. If the Agent were to employ a greedy strategy to maximize its reward at each timestep, it would result in short-term benefit but may not always lead to higher *returns*.

$$G(t) = R(t+1) + R(t+2) + \ldots + R(T) \tag{16.1}$$

where

T is the time at which episode ends

If you apply the same definition for *continuing* tasks, the *return* will become infinite, as reward will keep on accumulating forever. Discounting the rewards, as given by Eq. 16.2, eliminates this problem:

$$G(t) = R(t+1) + \gamma * R(t+2) + \gamma^2 R(t+3) + \gamma^3 R(t+4) + \ldots \tag{16.2}$$

where

γ is the *discount factor*, $0 \leq \gamma < 1$

Since γ is less than 1, raising it to higher powers decreases its value exponentially. Therefore future rewards contribute less and less percentage of their numeric value to the *return*, and this avoids the problem of infinite rewards. Such discounting is not needed in episodic tasks since the total reward per episode is bounded. Or, you could just use γ values close to 1.

There is another reason for *discounted rewards*. In Sect. 16.1.1.2, you had seen the concept of delayed rewards, i.e., the rewards being obtained now could be due to certain states in the past. However, intuitively, the impact of very distant states should be lesser. Alternately, the further out in the future you go, the lesser the impact of current state. Discounting reduces the contribution of future rewards that can be attributed to current state.

For stochastic environments, the rewards at each timestep can also be stochastic, which means $R(t)$ (the reward obtained at time t) can be a random variable. Therefore $G(t)$, the sum of random variables can also be a random variable. Therefore, the actual values of $G(t)$ can vary with episodes.

When rewards are stochastic, Agent should try to maximize *expectation* (i.e., average) of the return for an episode, i.e., maximize $E[G(0)]$. This can be achieved by maximizing $E[G(t)]$ for all values of time t. Maximizing expectation implies maximizing the average return over all episodes. The Agent should also minimize the *variance* of $G(0)$, as a secondary goal. High variance implies large variation in *returns* for episodes, making it difficult to estimate or predict an Agent's performance on a given episode.

Since *continuing tasks* are a single infinite episode, the Agent would have to maximize the *expected* (i.e., average) reward per timestep.

16.4 Agent's Behavior

The *return* $G(t)$ in an episode depends on the behavior of the Agent (which provides the actions) and not just the environment. The Agent's behavior is described by a function called the *policy network* or simply *policy*. It specifies the action to take for a given state.

When an Agent receives a state from environment, it looks up its *policy network* function to determine what action to take and passes that action to environment. This is called *following the policy* or executing the policy. Note that the *policy* does not consider current reward in determining future actions, since future actions are solely determined by current state. The current reward has already been obtained and so does not influence action from the current state.

Just as environment state transitions (and rewards) can be deterministic or stochastic (explained in Sects. 16.2.1 and 16.2.2), the *policy* can also be deterministic or stochastic. If the *policy* is deterministic, it specifies exactly one action for a given state. However, when the policy is stochastic, the actions form a probability distribution, and the Agent *samples* the distribution to determine a specific action. Since deterministic policy can be seen as a special case of stochastic policy, all discussions for stochastic policy would be applicable to deterministic policy. Hence, subsequent discussions will be limited to stochastic policy.

Let $\pi(a \mid s)$ denote *policy network*, a function that returns the probability of taking action a in state s. Since the probabilities of actions form a distribution, if $a_1, a_2, \ldots,$ a_N are the possible actions in state s, then $\pi(a_1 \mid s) + \pi(a_2 \mid s) + \ldots + \pi(a_N \mid s) = 1$.

The policy network can be stored as a lookup table that maps state-action pairs to probabilities. In a *parameterized policy network* described in Sect. 17.6.2, the lookup table is replaced by a *model* that approximates the policy network function.

Based on the understanding so far, the algorithm needs to find the optimal policy $\pi(a \mid s)$ that maximizes the expectation of returns (i.e., cumulative discounted future rewards) for each state.

Just like the environment's MDP, a stochastic policy network can be *stationary* or *nonstationary*, i.e., the probabilities of actions can either be constant over time or vary with time. A nonstationary policy can also be viewed as different stationary policies over time. The changes in policy with time are usually due to the Agent learning from experience.

16.5 Graphical Notation for a Trajectory

The (environment) state transition table and policy function can be represented in a single graph (actually a tree) to help visualize a *trajectory*. For the example graph shown in Fig. 16.3, the trajectory starts with state s_3, which is the root of the tree. The graph shows, for state s_3, the distribution of actions and their probabilities: (a_2, 0.4), (a_5, 0.1), (a_6, 0.5). The left most edge from action a_2 corresponds to MDP table entry: $\Pr(S_{t+1} = s_2, R_{t+1} = r_2 \mid S_t = s_3, A_t = a_2) = 0.4$. (notation explained in Sect. 16.2.2).

Consider an environment in state s_0 at time 0. The probability that it receives action a_0 is $\pi(a_0 \mid s_0)$. Once a_0 is received by the environment, the probability that it will transition to state s_1, with reward r_1, is $\Pr(s_1, r_1 \mid s_0, a_0)$. The product of these two terms represents the probability of transitioning from state s_0 to state s_1, i.e., $\pi(a_0 \mid s_0)$ * $\Pr(s_1, r_1 \mid s_0, a_0)$. Similarly, at time 1 (already in state s_1), the probability of transitioning to state s_2 is $\pi(a_1 \mid s_1)$ * $\Pr(s_2, r_2 \mid s_1, a_1)$.

Thus, the probability of transitioning from state s_0 to state s_2 is the product of the probabilities for transitioning from s_0 to s_1 and from s_1 to s_2.

Generalizing, given a *trajectory* $\tau = \{s_0\, a_0\, s_1\, r_1\, a_1\, s_2\, r_2 \ldots\}$, the probability of the trajectory $P(\tau)$ is the product of probabilities of all state transitions within the trajectory. For any trajectory, you can multiply the probabilities corresponding to the edges of the trajectory, and you will have the probability of that trajectory.

16.6 Value Function

Since you want to find an optimal policy, you need a measure of a policy's performance or goodness for an environment. One such measure is the *value function* of a policy. A *value function* can be a *state-value function* or an *action-value function*.

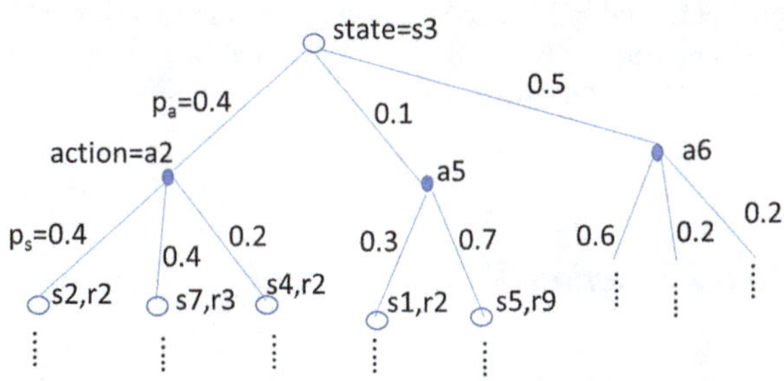

Fig. 16.3 Graphical representation of policy network and state transition function

16.6.1 State-Value Function

A *state-value function* assigns a numeric value (*state value*) to each state. The *state value* indicates how good a state is from an Agent's perspective. This lets you evaluate a policy. Loosely speaking, a policy that results in states with higher *state value* is good.

When both MDP and policy are deterministic, there is only a single *trajectory* starting from any state. So the value function for a state is simply defined as the *return* for that state, i.e., the cumulative reward received starting from that state until the end of the episode.

If the policy and/or environment is stochastic, there can be many trajectories starting from any given state. Therefore, the *return* from a given state is stochastic. The state's value is defined as the *expected return* an Agent receives by following its policy, starting from that state, as shown by Eq. 16.3.

The *state value*, i.e., value of a state s, for a policy π is denoted as $V_\pi(s)$. You can estimate $V_\pi(s)$ by averaging the actual *returns* over several (or all possible) trajectories starting from state s:

$$V_\pi(s) = E_\pi[G_t|S_t = s] \tag{16.3}$$

where

$V_\pi(s)$ is value of a state s for a policy π
E_π is the *expectation for policy* π
G_t *is the random variable for the return*
S_t is the state at time t

Since the policy controls the actions which in turn determine the transitions and rewards, *state values* are clearly a function of policy. So, *state values* are only defined for a policy. When comparing two policies, a policy π_1 is better than π_2 if the *state values* of all states in π_1 are higher than the corresponding *state values* in π_2.

It can be proven (proof omitted) that every MDP has an optimal policy π^*.Your goal is to find π^*. The corresponding value function v^* is the optimal value function. You can find π^* by finding v^*. It is good to know that such an optimal policy exists, even if the MDP itself is not known.

16.6.2 Action-Value Function

The *action-value function* is also called *Q-function*. The *Q-value*, denoted as $Q(s, a)$, is defined as the *expected return* starting from state s and taking action a, as given by Eq. 16.4:

$$Q_\pi(s,a) = E_\pi[G_t | S_t = s, A_t = a] \tag{16.4}$$

where

$Q_\pi(s,a)$ is Q-value corresponding to action a of state s for a policy π
E is the *expectation* for policy π
G_t is the random variable for *return*
S_t is the state at time t
A_t is the action at time t (in state s)

State-value functions and action-value functions are related by *Bellman expectation equation*, as shown by Eq. 16.5:

$$V_\pi(s) = E_a[Q_\pi(s,a)] \tag{16.5}$$

Intuitively, each state s has a valid set of actions and therefore is associated with a set of valid state-action pairs. If you average out the Q-values for all those state-action pairs, you get the *state value* for s.

Similar to optimal state-value function $v^*(s)$, you also have an optimal *Q-function* $Q^*(s, a)$. The *Bellman optimality equation* (Eq. 16.6) shows the relationship between optimal state-value functions and optimal action-value functions:

$$v^*(s) = \text{Max}_a Q^*(s,a) \tag{16.6}$$

where

Max_a denotes the maximum value of $Q^*(s, a)$ over all actions possible in that state

Intuitively, taking the max of *Q-values* instead of average seems appropriate since v^* will have highest possible state value.

Equations 16.7, 16.8, and 16.9 express Q^* in different forms. Convince yourself that these are correct representations. Equation 16.7 expresses action value as an *expectation* over the sum of reward and discounted next state's value since both rewards and next state values are stochastic. Equation 16.8 simply expands the expression for expectation by considering each of the N state-action pair combinations. Eq. 16.9 expresses the sum in $\sum$ notation and replaces state value by Q-value using Eq. 16.6 so that we get a relation between current and future Q-values:

$$Q^*(s,a) = E[r_{t+1} + \gamma v^*(s_{t+1})] \tag{16.7}$$

$$Q^*(s,a) = \Pr(s_1, r_1 | s, a)(r_1 + \gamma v^*(s_1)) + \Pr(s_2, r_2 | s, a)(r_2 + \gamma v^*(s_2)) + \ldots + \Pr(s_N, r_N | s, a)(r_N + \gamma v^*(s_N)) \tag{16.8}$$

$$Q^*(s,a) = \sum_{i=1}^{N} \Pr(s_i, r_i | s, a)\, (r_i + \gamma \text{Max}_a Q^*(s_i, a)) \tag{16.9}$$

Section 16.7.1.2 will introduce ways to estimate *Q-function* even when MDP is not known, as is common with many reinforcement learning problems. This is a clear advantage of *Q-functions* over *state-value functions*.

16.7 Methods for Finding Optimal Policies

The methods for finding optimal policy for reinforcement learning problems depend on a few factors, such as:

- How much of an MDP model is known
- Whether they are *model-based* or *model-free*
- Whether they are *on-policy* or *off-policy*

16.7.1 Agent's Awareness of MDP

An MDP model of *environment* contains five ingredients:

1. S is the state space, i.e., the set of all states.
2. A is the action space, i.e., the set of all actions.
3. Pr is the probability transition function, i.e., a function specifying the probability of each possible state transition.
4. R is the reward function, i.e., a function specifying the reward for each possible state transition
5. γ is the discount factor, a positive real number less than or equal to 1

In any reinforcement learning setting, Agent's awareness of an *environment* can be categorized into three types – MDP known, MDP unknown, and MDP partially known.

16.7.1.1 MDP Known

The Agent has full knowledge of MDP model. For example, in a chess game, it is clear how the state (board configuration) changes with action (making a move). If MDP is known, and the number of states and actions are limited, it is possible to compute the optimal policy using iterative methods or linear programming methods. These methods take in the MDP model and output the optimal policy. Section 16.8 describes an iterative method to obtain optimal policy. Such iterative methods can compute the optimal policy even before Agent starts interacting with the environment, so they are known as *planning* methods and not learning methods.

If there are a large number of states, precomputing the policy may not be possible. In such cases *sampling* and *rollout*-based methods are used. You will see some of these methods in the next chapter.

16.7.1.2 MDP Unknown

MDP being unknown implies that Agent is aware of MDP's existence but does not know its state transition table and rewards. The set of valid states and actions are always known, as they are part of the problem definition and required for an Agent to communicate with the environment. For example, a robot may be put into a new environment with no prior knowledge of the environment. In this situation, the Robot Agent must know the set of valid states and actions but doesn't have to know the state transition function and the rewards.

There are three basic approaches to finding optimal policy when MDP is not known. First approach is to learn the MDP model itself while interacting with the environment. One example of this method is Dyna2 (not covered in this book). Second approach is to learn the value function while interacting with environment and derive the optimal policy from it. Next chapter describes two such methods: *Monte Carlo* learning and *temporal difference learning* or *TD learning*. The third approach is to ignore the MDP model and directly optimize the policy based on experience gained from interacting with environment. Next chapter will describe the *policy gradient* method that follows this approach.

In all of these approaches the Agent has to learn the optimal policy while interacting with the environment. In other words, Agent has to learn the optimal policy from experience (either its own experience or some other agent's experience).

Every time an Agent observes a new state transition and reward as a result of its action, it discovers (or learns) a new entry in the MDP table (excluding the state transition probability value). The Agent is effectively *sampling* the environment (i.e., its MDP) to learn about it. Therefore, these observations are known as *samples*. A trajectory is also a sample.

For the purpose of this discussion, an *experience* is a set of *samples* where each *sample* has just one state transition and reward. In other words *experience* is the set of unordered and possibly unrelated tuples: (s_0, a_0, s_{n0}, r_0), (s_1, a_1, s_{n1}, r_1), ..., (s_N, a_N, s_{nN}, r_N) where the *i*th sample consists of state transition from state s_i to state s_{ni}, caused by action a_i, leading to reward r_i.

Both Monte Carlo and TD learning involve learning of value functions by the Agent via *sampling*. In Monte Carlo approach, the Agent accumulates a large number of episodes and then computes the value function at the end based on all the episodes. In TD learning, the Agent starts with an initial estimate of value function (say all states have zero values) and then updates the estimates every time it samples the environment.

16.7.1.3 MDP Partially Known

An example of a partially known MDP is in the game of poker where the Agent only knows some of the cards and hence has partial information about the environment state.

The Agent is assumed to know the set of valid states and actions, but only has partial information on state transition probabilities and rewards.

When MDP is only partially known, the optimal policy can be computed by model-free reinforcement learning algorithms since they don't need an MDP. In practice these algorithms work well when using *function approximation* (explained in Sect. 17.5) to model the states, since function approximations allow you to generalize from known states to unknown states. Next chapter describes some of these methods.

Such environments can also be modeled by different types of MDPs known as POMDPs (partially observable MDPs). There are specialized methods to solve for POMDPs which are beyond the scope of this book.

16.7.2 Model-Based and Model-Free Reinforcement Learning

The approaches that make use of MDP model (i.e., model is used or learnt by Agent while interacting with environment) are known as *model-based reinforcement learning* algorithms. These approaches generally compute the value function from the MDP model, and the value function is then used to compute the optimal policy.

The approaches that do not make use of MDP model are known as *model-free reinforcement learning* algorithms. *Model-free learning* applies to many real-world situations where the MDP is either not known or is hard to model.

16.7.3 On-Policy and Off-Policy Reinforcement Learning

In *on-policy* learning, the Agent learns solely by following its current policy and observing the state transitions and rewards firsthand, i.e., it learns by direct *experience*. *Off-policy* learning on the other hand refers to the Agent learning without following its own policy. Instead it learns by observing other's experience. The observations could be from another Agent that is following its own policy, or it could be from its own or other Agent's past *experience* (typically stored in a database).

16.8 Policy Iteration Method for Optimal Policy

This is an iterative procedure that starts with a random policy and gradually improves it in each iteration based on a value function. The iterations continue until the values converge, at which point you have an optimal policy.

16.8.1 Computing Q-function for a Given Policy

Start off with zero (or, very small) Q-values. Use Eq. 16.9 for computing and updating *Q-values* for current state-action pair using Q-values from future state-action pairs. As an example, Eq. 16.10 shows the Q-value computation, for a hypothetical MDP that has only two-state transitions from state pair (s_0, a_0) – transition to state s_1 with reward r_1, with probability 0.4, and transition to state s_3 with reward r_3, with probability 0.6:

$$Q(s_0, a_0) = 0.4(r_1 + \gamma \text{Max}_a Q(s_1, a)) + 0.6(r_3 + \gamma \text{Max}_a Q(s_3, a)) \tag{16.10}$$

You can use Eq. 16.9 to iteratively update all Q-values – updating one value at a time. Iterate enough number of times. Don't worry yet about convergence.

16.8.2 Policy Iteration

For any state s, you have a set of *Q-values*, corresponding to actions in that state. Define your policy to pick the action corresponding to the maximum *Q-value* among the possible actions. This results in *greedy policy*. The following pseudo-code provides the algorithm for iterating over the policy till you get optimal solution.

```
Set π to a random policy
Let prevQF and curQF store the Q-value function
Initialize them with all zero values
Loop forever
      Compute Q-value function for policy π and store in curQF
      Compute a new greedy policy from curQF
      set π to the new greedy policy
      if  significant improvement in Q-values
          set prevQF = curQF
    else
          exit loop // convergence achieved
End loop forever
```

Alternately, you can use *state values* instead of *Q-values* for policy iterations.

Now that you have most of the building blocks needed for reinforcement learning, the next chapter talks about the algorithms in use for situations, when MDP is not known.

Chapter 17
Reinforcement Learning Algorithms

In the previous chapter, you were introduced to major aspects of reinforcement learning. This chapter takes you through the next steps on dealing with those challenges in order to form the algorithms for reinforcement learning. Reinforcement learning is an area of very active research, and new variations of algorithms are proposed regularly. An understanding of this chapter will provide you with a good basis, so that you can appreciate not just the current generation algorithms but also understand new research findings in this area. RL involves interaction of multiple components and concepts, which you have already seen in prior chapters in this book.

17.1 Monte Carlo Learning

Monte Carlo learning is one of the methods used to estimate the *value function* (i.e., the set of state values) when Markov decision process (MDP) is unknown (Sect. 16.7.1.2). The Agent has to learn the MDP based on its observations of the environment over several episodes. The idea is to estimate the state values for a policy by following the policy for many episodes, computing *the return* for each state in each episode, and then averaging the *returns* for each state over all the episodes. This mechanism provides a good estimate of state values *expectations*. The accuracy of the estimates increases with number of episodes. In practice, estimates for state values are obtained from 1000s to millions of episodes depending on the nature of the RL problem. This method only applies to episodes.

G. Rebala et al., *An Introduction to Machine Learning*,
https://doi.org/10.1007/978-3-030-15729-6_17

17.1.1 State Value Estimation

It is easier to understand the estimation process using an example. For ease in understanding, consider only two episodes. In reality the Agent requires many episodes to learn.

Tables 17.1 and 17.2 represent the two episodes. The episodes are represented in the first three columns. The last column labelled *return* is not part of the episode. Using Table 17.1 to understand the table entries, the first row says that, at time $T = 1$, environment is in state s_5 and it received action a_1. The second row says that, at time $T = 2$, environment has transitioned to state s_4, received action a_3, and issued reward r_4. In the last column, *return* is computed once, only after the terminal state is reached, based on the description in Sect. 16.3, assuming discounting factor γ to be 1.0.

Looking at the two episodes, the estimated state value for s_4 (i.e., $V_\pi(s_4)$) is average return for the state over the two episodes, viz., $((r_8 + r_1 + r_5) + (r_9 + r_{17} + r_5 + r_2))/2$. For estimating $V_\pi(s_5)$, only Table 17.1 contributes, and the value is determined to be $r_4 + r_8 + r_1 + r_5$. So, you can estimate the state values for each of the state observed in all the episodes.

Notice that state s_9 occurs twice in Table 17.2. And, the *return* values are different for the two entries. You could now decide to consider the return value from only one of the entries. For example, in *first-visit Monte Carlo*, you consider the first occurrence of s_9, thus, considering the return value as $r_{17} + r_5 + r_2$. Or, you could consider

Table 17.1 Episode 1 for MDP learning

Time T	State-Action Pair	Reward	Return
1	s_5, a_1		$r_4 + r_8 + r_1 + r_5$
2	s_4, a_3	r_4	$r_8 + r_1 + r_5$
3	s_8, a_3	r_8	$r_1 + r_5$
4	s_9, a_7	r_1	r_5
5	s_{11} (terminal state)	r_5	

Table 17.2 Episode 2 for MDP learning

Time T	State-Action Pair	Reward	Return
1	s_4, a_3		$r_9 + r_{17} + r_5 + r_2$
2	s_9, a_8	r_9	$r_{17} + r_5 + r_2$
3	s_{17}, a_4	r_{17}	$r_5 + r_2$
4	s_9, a_5	r_5	r_2
5	s_{11} (terminal state)	r_2	

each entry for s_9 (called *every-visit Monte Carlo*) to get two different entries for *reward* for s_9. Depending on which method you use, the *reward* values from either just the first entry from Table 17.2 or both entries from Table 17.2 will contribute in estimating the state value for s_9, in addition to the reward values from Table 17.1.

Estimating state values for an unknown MDP may be the simplest way to evaluate a policy. However, it has limited use since computing an optimal policy based on state value function requires knowledge of state transitions. Estimating state-action values, or Q-values is more useful since they can be used to derive the optimal policy without knowing the MDP.

17.1.2 Action Value Estimation

The *Q-values* for state-action pairs are computed in the same way as for state value function, except that the value is assigned to state action pair. Once again, considering the two episodes shown in Tables 17.1 and 17.2, $Q(s_4, a_3) = ((r_8 + r_1 + r_5) + (r_9 + r_{17} + r_5 + r_2))/2$. Multiple occurrences of state-action pairs in an episode are handled in the same way as multiple occurrences of states in state value function computation.

Computing the mean of state-action values after all episodes have been completed requires storing all the episodes before computing the mean, which will need a huge memory for large number of episodes. It is much more efficient to maintain the *running average* for each state-action value and update it after each episode.

This method of estimating value functions can also be used when MDP is known, but is so large that value iteration and other methods that depend on storing the MDP in memory become impractical.

The Monte Carlo methods have following drawbacks:

- Requires waiting for the whole episode to end before computing the values, since the methods depend on the *returns* for states, which is only known at the end of the episode.
- The methods don't apply to *continuing tasks*, since there is no end to an episode.
- The *return* is a sum of (discounted) rewards, where each reward is obtained stochastically. Therefore, the sum of rewards tends to have high variance (i.e., the actual *returns* will vary a lot between episodes) which is undesirable.

The above issues are addressed by temporal difference learning methods, or TD learning methods for short.

17.2 Estimating Action Values with TD Learning

TD methods update *Q-value* estimates at every timestep (or after a fixed number of timesteps). So:

- You don't have to wait for the episode to finish.
- Thus, continuing tasks are equally good for these methods.
- Variance is reduced since at every timestep the update involves just one random variable instead of many random variables.

The TD methods start off with some initial values for all state-actions (say zeros) and then iteratively update the estimates after a fixed number of steps. TD(n) represents a method where estimates are updated after every n timesteps. Based on Bellman update, given by Eq. 16.9, an updated Q-value can be computed by Eq. 17.1:

$$Q\mathrm{new}_{t+1} = r_{t+1} + \gamma Q(s_{t+1}, a_{t+1}) \tag{17.1}$$

where:

$Q\mathrm{new}_{t+1}$ represents the updated *Q-value* at time $t + 1$, corresponding to state and action (s_t, a_t) at time t.
r_{t+1} represents reward at time $t + 1$.
γ represents the discounting factor.
$Q(s_{t+1}, a_{t+1})$ represents the current *Q-value* of the state-action pair at time $t + 1$.

Since the new *Q-value* estimates are based on existing *Q-value* estimates, we refer to this method as *bootstrapping*. Intuitively, this new estimate should be slightly more accurate than previous estimate because it is now based on an actual reward value (r_{t+1}) and an actual transition (s_t to s_{t+1}). However, notice that its value is based on another estimate, $Q(s_{t+1}, a_{t+1})$, so it is a *biased estimate*, unlike Q-value estimates from Monte Carlo methods which are *unbiased estimates*.

Since the environment may be *nonstationary*, you need to lessen the influence of older rewards. This can be implemented by introducing another factor α, similar to *learning rate*, and is given by Eq. 17.2, where only a portion of the update is actually applied. So, instead of changing Q-value to the new value, you update Q-value to approach closer to the desired new value. Replacing $Q\mathrm{new}_{t+1}$ using Eq. 17.1, you will get Eq. 17.3:

$$Q(s_t, a_t) = Q(s_t, a_t) + \alpha(Q\mathrm{new}_{t+1} - Q(s_t, a_t)) \tag{17.2}$$

$$Q(s_t, a_t) = Q(s_t, a_t) + \alpha(r_{t+1} + \gamma Q(s_{t+1}, a_{t+1}) - Q(s_t, a_t)) \tag{17.3}$$

where:

α represents a fraction of the computed update

The pseudo code for TD(0) method is given as:

Initialize Q(s,a) to zero for all possible (s,a) pairs
Foreach episode
    Foreach time t = 0, 1, 2, ... (until episode ends)
        Collect one experience by following current policy.

Update $Q(s_t, a_t)$ using equation 17.3
end foreach time
end foreach episode

If you notice carefully, the *Q-value* updates are somewhat similar to θ updates in Linear Regression or weights update in neural networks. You start with random choice of *Q-values*. With each observation, you determine whether the value should be increased or decreased; you also get a sense of by how much and then *adjust* the value by a small amount. You move *toward* the new desired value, not *exactly to* the new value calculated/estimated.

17.3 Exploration vs Exploitation Trade-Off

To better understand this trade-off, consider a situation where you want to buy clothes. If you are new to the place, you have to *explore* a few stores first to see which ones you like. On the other hand, if you are already familiar with the shops in that area, you would *exploit* the information you already have. And, sometimes, you may *explore* even more stores, in hope of finding even better choices. But, you will *explore* less often, as you get more familiar with the stores.

Coming back to the current problem: when the Agent starts off, it has to *explore* the environment states by picking actions randomly, so that it can reach a variety of states (or state-action pairs) and discover high-value trajectories. Later on when it has enough information, it can pick actions more purposefully to visit known high-value states or trajectories to maximize its rewards. So, the Agent can start off with a stochastic policy, i.e., one that selects random actions, and slowly decrease the randomness over time so that eventually it has a good deterministic policy.

17.3.1 ϵ-greedy Policy

An *ϵ-greedy* policy is a stochastic policy derived from a deterministic policy in order to encourage *exploration*. It is composed of two policies:

- A deterministic policy
- A random policy (i.e., actions are picked uniformly at random)

An Agent chooses the random policy with probability ϵ and (therefore) the deterministic policy with probability $1 - \epsilon$. Then the Agent chooses the action as per the selected policy. Keep a high value of ϵ as you start, so that there is more *exploration*. ϵ can be steadily decreased over time so that Agent picks deterministic policy more often to use *exploitation*.

In general, algorithms that iteratively improve a deterministic policy are better off following the ϵ-*greedy* policy derived from the current deterministic policy in each iteration in order to encourage *exploration*.

17.4 *Q*-learning

Q-learning is an *off-policy* (Sect. 16.7.3) learning method. It learns the optimal greedy policy from *Q*-function. The key feature of this algorithm is that it can learn the optimal *Q*-function from any available *experience* for that environment. For example, it can learn from past *trajectories* followed by other Agents or policies. With good available *experience* (i.e., covering most of the valuable state-action pairs and state transitions), the learning can converge to optimal *Q*-function and hence find an optional policy.

The idea is that at any time, you have some estimates of *Q*-function (including random *Q*-function at start) and you can update the estimates to make them more accurate by incorporating actual MDP data from each sample in an *experience*. A sample of *experience* is a 4-tuple from following some policy, represented as (s_t, a_t, s_{t+1}, r_{t+1}), which describes a state transition: state s_t with action a_t, transitioned to state s_{t+1} with reward r_{t+1}. This same information would be a row in *MDP table*, without the probability value. The *Q*-value update is given by Eq. 17.4:

$$Q(s_t, a_t) = Q(s_t, a_t) + \alpha(r_{t+1} + \gamma(Q\max(s_{t+1})) - Q(s_t, a_t)) \tag{17.4}$$

where:

$Q\max(s_{t+1}) = \max_{a'} Q(s(t+1), a')$ is the maximum of all *Q-values* associated with state s_{t+1}.

And, other symbols have the same meaning as in Eq. 17.3.

The expression $r_{t+1} + \gamma(Q\max(s_{t+1}))$ is called the *Q-learning target* since $Q(s_t, a_t)$ is being updated toward that value. The reason this algorithm is *off-policy* is that the *Q-value* update includes the max of *Q-values* for a state, which means the updates are not directly determined by the policy.

The *Q*-learning algorithm starts with some initial *Q-function*, derives a greedy policy for it, and then improves both the greedy policy and the *Q- function* in an iterative fashion. In any iteration, there are two policies in use:

- The greedy policy known as the *target policy*. This is the policy you are trying to improve. This policy is not followed by the Agent.
- An ϵ-*greedy policy* (see Sect. 17.3.1) created from the *target policy*. This policy is called *behavioral policy*, and it is followed by the Agent. In general, this policy could be any stochastic policy that ensures enough *exploration*, to help *Q-values* converge to optimal values.

The pseudo code for the Q-learning algorithm where the *experience* is generated using a ϵ-greedy policy is given below:

Initialize:
Q-values to zeros
behavior policy to a random policy
set α to a reasonable value // end initialization
Start a new episode or trajectory in a random initial state.
Repeat until Q-values converge
Follow behavior policy for one step to generate the 4-tuple: [s(t), a(t), s(t+1), r(t+1)]
Update Q-value using equation 17.4
Create greedy policy PG from current Q-value function
Replace behavior policy with ϵ-greedy policy derived from PG
If enough timesteps have passed since last ϵ update, then reduce ϵ by a small amount
If an episode ends then start a new episode with a random initial state
End repeat till convergence

17.5 Scaling Through Function Approximation

In many RL problems, the number of states can be very large. For example, the board game *Go* has about 10^{170} states, which is practically infinite. RL problems with very high number of states exhibit two major challenges:

- You cannot store the MDP table or the computed *value function* in memory.
- It is almost impossible to explore even a tiny fraction of possible states or state-action pairs; which in turn means you cannot estimate the *return* (or, goodness) of a state.

The model-free algorithms don't require an MDP. Therefore we only care about storing the value function (usually the state-action function).

Both the challenges can be addressed by *function approximation*. In earlier chapters, you have learnt about Linear Regressions and neural networks, which can approximate any function given enough samples of inputs and outputs. You can use a deep neural network to represent the value function. This solves both the challenges. First challenge is solved since value function can be represented compactly in memory. Second challenge is solved because the value function can *generalize* to unknown states and actions. That is, it is able to give an estimate of values even for states and actions that are not visited.

17.5.1 Approximating the Q-function in Q-learning

In *Q*-learning, each state transition results in an *experience* (i.e., a *trajectory* with just one transition) and the associated *Q-learning target* through Eq. 17.4. This provides you with the following labelled example for training the neural network:

Input: $s(t)$, $a(t)$
Output: *Q-learning target*

The model is denoted $Q(s,a; \theta)$ where θ is the set of parameters (or weights in the neural network). This is a regression problem and not a classification problem since the output is a real value.

You use mini-batch training in an iterative loop. In each iteration, generate N experiences, resulting in N labelled training examples, and use them to update your parameters. Continue to iterate until parameters converge.

One of the major drawbacks of this method is that the successive *experiences* are not independent of each other. They are *correlated* by definition since the states and actions from the past determine the next set of states and actions. This is a problem for any supervised model, since the training only works well if the individual training examples are independent of each other. The other disadvantage is that since both state and actions are inputs to the neural network, if a state has N possible actions, then you have to do N forward traversals of neural network to get the N probabilities, thus slowing down the overall computation.

DQN (*Deep Q-network*) is a deep neural network with a variation of *Q-learning*. It was used to successfully train an Agent to play different Atari video games at expert level, purely based on reading the game state from the raw pixels from the screen. It solved the two problems mentioned in the previous para.

- Its neural network has just state as an input, but N outputs, one for each action. This means only one forward pass is needed to get all the action probabilities, thus making it much faster.
- It used *experience replay* to minimize the problem of correlated samples. This involves storing the *experiences* from multiple episodes in a database and choosing a random subset of *experiences* from the database, to be used as a mini-batch for training the neural network using stochastic Mini Batch Gradient Descent. As an additional benefit, the same database can be sampled several times to make more efficient use of available training data.

17.6 Policy-Based Methods

So far you have seen methods that find optimal policy by first computing a *value function* (such as *Q-value* function) and then deriving the policy from that value function. Policy-based methods on the other hand compute a policy directly without the intermediate step of computing a *value function*. Most of these methods are based on computing gradients so they are commonly called *policy gradient* methods.

17.6.1 Advantages of Policy Gradient Methods

Policy gradient methods can produce stochastic policies, unlike value-based methods such as Q-learning which only produce deterministic policies (or near-deterministic policies such as ϵ-*greedy* policies). This is an advantage because of two reasons. First, for some problems the optimal policies are stochastic. A classic example is the *rock-paper-scissors* game whose optimal policy is stochastic (pick one of the three items uniformly at random). Another example is *Poker* where the best strategy is stochastic (whether to bluff or not is best decided stochastically). The second reason is that they can handle incomplete state information. When state information is incomplete, the optimal deterministic actions, if any, are not known for all states, so the next best option is to pick one stochastically. A good example is *Poker* where the Agent can't see all the cards, so has incomplete state information. More generally, if an Agent can't choose between a set of actions because the state information is incomplete or noisy, then the best option is to choose one of those actions stochastically.

Policy-based methods tend to be simpler for certain problems compared to value-based methods such as Q-learning. Policy-based methods are guaranteed to converge to local maximum, whereas there are no such guarantees for value-based model-free methods such as Q-learning.

17.6.2 Parameterized Policy

Policy gradient methods require a parameterized policy $\pi_\theta(s,a)$, where θ is a set of parameters, such that varying the parameters gives us different policies, and you can find the parameters for the best policy through supervised learning. Recall that a policy is a function whose input is the state vector and outputs are probability values for possible actions. So you can model the policy as a supervised model (Logistic Regression or neural network) for classification with a *softmax* output layer.

For complex states, such as for images from video games or board games, the raw state (pixels) is usually fed into a *CNN* (covered in Chap. 15), and its output becomes the state vector that usually feeds into a single fully connected layer with one node per action, which then feeds the *softmax* layer.

17.6.3 Training the Model

For training the model, you use some measure of how good a state-action pair is. Using this measure, a high-level algorithm for training would be:

Initialize parameters of policy network to random values
Loop forever
Follow the policy network to generate multiple episodes (or trajectories)
From these episodes identify the good and bad state-action pairs
Adjust the policy network parameters such that the network is more likely to prefer the good state-action pairs, and less likely to prefer bad state-action pairs.
if parameters converge then exit the loop
end loop

For the above algorithm, you still need to know which state-action pairs in a trajectory are good and which are bad. Also, you need an appropriate measure, whose gradient you can use to adjust parameters in the right direction and right quantity.

Identifying good and bad states is the *credit assignment* problem (mentioned in Sect. 16.1.1). Intuitively, state-action pairs with high *Q-values* are good, and those with low *Q-value* are bad. Assuming you have an estimate of *Q-values*, you can adjust the parameters in the direction of the policy gradient, with strength proportional to the *Q-value*. If *Q*-value is positive, then parameters are adjusted in the direction of gradient (called *gradient ascent*), and if *Q*-value is negative, then parameters are adjusted in the opposite direction of the gradient (called *gradient descent*). The gradient (of the policy) at the current state-action pair is given by $Q(s, a) \nabla \pi(s, a; \theta)$, where, $\nabla\pi$ denotes derivative of the policy π. For ease of computation, use $\nabla \log \pi$, instead of $\nabla\pi$. Using this value for gradient, Eq. 17.5 gives the equation for updating the parameters:

$$\theta = \theta + \alpha\, Q(s,a) \nabla \log\pi(s,a;\theta) \tag{17.5}$$

where:

α is the usual *learning rate* from supervised learning

The *policy gradient theorem* given in Eq. 17.6 (proof omitted) validates our intuition behind Eq. 17.5:

$$\nabla J(\theta) = E\left[Q(s,a) \nabla \log\pi(s,a;\theta)\right] \tag{17.6}$$

where:

$\nabla J(\theta)$ represents gradient of an objective function
E is the *expectation*

The objective function $J(\theta)$ for episodic tasks is the *return* $R(0)$, while for continuing tasks, it can be the average reward per timestep.

The pseudo code for policy gradient methods is given below:

Initialize θ with random values
loop forever
generate episode E = s_0, a_0, r_1, s_1, a_1, . . ., s_{N-1}
for t = 0 to N-1
update θ as per equation 17.5
end for
if θ converges then exit forever loop
End loop //forever

You will get different policy gradient methods depending on how you estimate $Q(s,a)$.

17.6.4 Monte Carlo Gradient Methods

The simplest estimate for $Q(s,a)$ is the *return* of the episode, $G(0)$. With this the parameter update equation given by 17.5 is simplified to Eq. 17.7:

$$\theta = \theta + \alpha G(0) \pi(s_t, a_t; \theta) \tag{17.7}$$

This update effectively says that all state-action pairs in the episode are equally good (if $G(0) > 0$) or equally bad ($G(0) < 0$). This is a crude estimate since in reality some state-action pairs in an episode are usually better than others. This oversimplification leads to high variability in episode *returns*. A better estimate for $Q(s,a)$ is to use the *episodic return* $G(t)$, which considers discounted rewards starting at time $t + 1$, until the end of episode. That would modify the update equation to 17.8:

$$\theta = \theta + \alpha G(t) \pi(s_t, a_t; \theta) \tag{17.8}$$

Methods based on using *returns* as estimates of *Q-values* are known as a Monte Carlo gradient methods since you are sampling from episodes. They are more commonly grouped together as *REINFORCE* algorithm.

17.6.5 Actor-Critic Methods

Alternately, you could compute *Q-value* estimates, as explained in Sect. 17.4, and use them just for gradient estimates of policy parameters. The policy gradient methods that use *Q-values* for gradients are called *actor-critic* methods. *Critic* refers to the portion of the method that estimates *Q-values*, and *actor* refers to the portion of program that updates the gradient based on *Q-values* provided by *critic*.

17.6.6 Reducing Variability in Gradient Methods

Reducing variability in gradient values collected at each timestep is the biggest challenge in policy gradient methods since it directly relates to the *credit assignment problem* and the associated *sampling problem* (need for too many samples to learn a good policy).

If you knew exactly which state-actions are good and by what magnitude, then you would have minimal variance. Lacking that information leads to overshooting or undershooting the gradient values. This in turn leads to large variation in policy during parameter updates, causing slower convergence, which in turn requires us to collect more gradient samples, leading to the *sampling problem.*

For example, consider a sequence of ten positive *Q-values* for a trajectory: 10, 9, 8, 7, 6, . . ., 1. You can see clearly that the *Q-values* are trending downwards. However since they are all positive, the current methods of gradient updates would consider all actions along that trajectory as good, when in reality you probably should not encourage actions which result in *Q*-values that are below a running average (say, of last ten *Q-values*). It is more appropriate to maintain separate running averages for each state since gradients are computed for a state-action pair.

This issue manifests as high variance in sampled gradient values. One way to address this issue is to subtract a running average of *Q-values* from each estimated *Q-value*. The value being subtracted is called a *baseline function*. Subtracting the baseline function from a *Q-value* does not change the *expectation* of the gradient (proof omitted). With this concept of baseline function being subtracted, the parameter update equation is modified to Eq. 17.9:

$$\theta = \theta + \alpha(G(t) - b(t))\pi(s_t, a_t; \theta) \tag{17.9}$$

where:

b(*t*) refers to the baseline function. It's a function of time, since it keeps running average

G(*t*) refers to estimated *Q*-value, though you can use computed *Q-value* as well

State value function V(*s*) (explained in Sect. 16.6.1) can stand for *b*(*t*) for each state. Using this replacement, the update equation is further modified to Eq. 17.10:

$$\theta = \theta + \alpha(G(t) - V(s_t))\pi(s_t, a_t; \theta) \tag{17.10}$$

17.7 Simulation-Based Learning

Policy iteration (explained in Sect. 16.8) is one of the methods to learn the optimal policy, without even interacting with the environment. However, this method only works if the state space is not very large. In many real-world applications, the state space is too large. In such cases, you can benefit by simulating the environment. That is, the Agent can embed a program within itself that behaves like the environment. That program would take state and action as input from Agent, and output the reward and new state to the Agent, all with no involvement from the real environment.

For example, suppose an Agent is learning to play chess. The environment is the chess board and its configuration. Chess moves cause the state changes. Winning, losing, or drawing games results in rewards.

In a real game against a human, the Agent would pass its chess move as an output on a terminal, and a human would make that move on the board. Then when the human opponent makes a move, someone would enter the new state to a terminal to be read by an Agent. For a more automated interface, you could have a robotic hand controlled by Agent make the chess moves and have a camera to help a robotic hand make the chess moves and capture the chess board configuration. This robotic and camera automation can be implemented as supervised models which will not be described here.

Now, instead of real game, think of a program to simulate the environment. The program would encode the chess board configuration in some data structure and code the rules of chess game. When the program is called to make a move, it would simply alter the board configuration as per the move (of course after checking that it's a legal move) and return the new board configuration to the caller.

Such simulated environments allow the Agents to plan ahead and optimize their policies before even interacting with the environment. One way to plan ahead is to simply interact with the simulated environment over several episodes to derive an optimized policy, very much like the *model-free reinforcement learning* methods, such as *Monte Carlo* and *Q-learning* methods. The Agent is said to be learning from *simulated experience*.

The Agent can learn an optimized policy offline by interacting with simulated environment. However learning an optimized policy for a large state space may still be hard. It may require a large number of episodes to reach sufficiently large number of states and state-action pairs. This challenge can remain even after using function approximation (Sect. 17.5) to generalize the *value functions* for unvisited states or state-action pairs.

There is another way to learn more efficiently offline – Agent can learn through *simulation* while interacting with the real environment.

During the interaction with the real environment, the Agent can learn the best action for the *current state* by interacting with the *simulated* environment starting from *current state* and cycling through all actions for the current state, for several simulated episodes. Each simulated episode would start with a fixed state (current state of real environment) and action, and follow a random simulation policy so that it can explore different trajectories. Once the Agent has sufficient number of simulation *experiences* from the *generated* episodes, it can compute the value function to pick the best action for the current state of the real environment.

This reduces the complexity of the problem since you are not exploring the whole state space. You only care about the portion of MDP that comes into play from the current state. The states and actions not reachable from the current state become completely irrelevant. The Agent is said to be engaged in *decision time planning* since the Agent is planning ahead just before deciding on an action. The Agent can keep generating simulated episodes until it needs to respond to real environment. For example, if the Agent has 3 minutes to make a chess move, then it can simulate for 3 minutes.

Suppose the environment is in state s, at time T in an episode. The Agent has to generate simulated episodes for every possible action starting from state s, in order to select the best action. The pseudo code for Agent's interaction with simulated environment at time T is given below:

```
Loop until it is time to respond to real environment
Foreach valid action a_c from state s
        Start a new simulated episode with state s'=s, and, an action a'=a_c
  Loop until reaching a terminal state
            get new state s_new and reward r from the simulated environment
            update simulation Q-values with the new experience
            select next action a_new for s' by following the simulation policy
            set s' to s_new, a' to a_new
     End-loop // terminal state reached
          Update simulation Q-values if rewards are issued on terminal states
  End-loop // end simulated episode for action a_c
End-loop // end all simulated episodes
Pick the best action for the real environment based on Q-values from simulation
```

The whole loop over multiple episodes is called a *round of simulation*. The last step is for the Agent to pick the best action based on the *Q-values* of state-action pairs, corresponding to the current state of the real environment. Once the action is passed to real environment, the Agent gets a new state (after opponent's move) and reward, the time is incremented, and the Agent can do another *round of simulation*, starting from the new state. The Agent can perform these simulations for every timestep of real episode.

You get different simulation-based methods depending on how you estimate $Q(s, a)$ and how you handle the simulation policy π_{s}. The simplest method is based on *Monte Carlo Q-value* estimation based on average returns, as described in Sect. 17.1. The simulation policy is just a policy that picks actions uniformly at random. Generating simulated episodes using random policy is typically referred to as *Monte Carlo rollouts* or simply *rollouts*.

In games the rewards are typically only given at the end, usually a simple +1 for a win, 0 for a draw, and −1 for a loss. The rollouts are also called *playouts* since the game is played out until the end using random moves. Since each rollout results in just one reward at the end of the episode, you can skip the *Q-value* computation and estimate the value of state-action pair by counting the number of wins and losses in a round of simulation and taking the win/loss ratio as the value.

17.8 Monte Carlo Tree Search (MCTS)

The Monte Carlo Tree Search (MCTS) applies to Agent's interaction with *simulated* environment. It is a popular method used in game play that generated a lot of excitement when it was used to vastly improve the performance of board game *Go*. It has been researched for many years resulting in several variations. This section describes the standard version.

MCTS has been mostly applied to games where there's a single reward at the end of a game. Therefore the discussion here applies to episodes where there is just one reward at the end of an episode (usually +1 for a win, −1 for a loss, and 0 for a draw), unless otherwise noted.

In the simple Monte Carlo method, you discard a simulated episode after you record the reward from it. On the other hand, in *MCTS* you store all the simulated episodes for a *round* of simulation in a *search* tree (in memory). A *round of simulation* is just a set of simulated episodes.

17.8.1 Search Tree

The search tree is the usual graph tree data structure. Search tree is central to MCTS, so is explained here briefly.

A search tree can compactly store multiple sequences of some elements such that you can quickly check if a given sequence matches any of the sequences in the tree. For example, suppose you have these three sequences of length four: (a_0, a_1, a_3, a_4), (a_0, a_1, a_3, a_6), and (a_0, a_2, a_3, a_2). A search tree for storing these three sequences of data is shown in Fig. 17.1.

Now, given a sequence $S = (s[0], s[1], s[2], s[3])$, you start by comparing $s[0]$ with the root node of the tree (labelled a_0). If it matches then you walk down the tree one level to check if any child of a_0 matches $s[1]$. Suppose node a_1 matches $s[1]$. Then walk down one more level to see if any of a_1's child nodes match $s[2]$ and so on. If at any time you can't find a matching node, then you conclude the sequence S does not occur in that tree. This method also helps you determine if there is a partial match.

To build such a search tree from a set of sequences, just follow the same procedure as above for each sequence, except if you don't find a matching sequence; then add appropriate nodes and edges so that the sequence becomes part of the search tree.

For the same example, start with an empty tree. After processing the first sequence, the tree consists of four nodes and four edges with no branches:

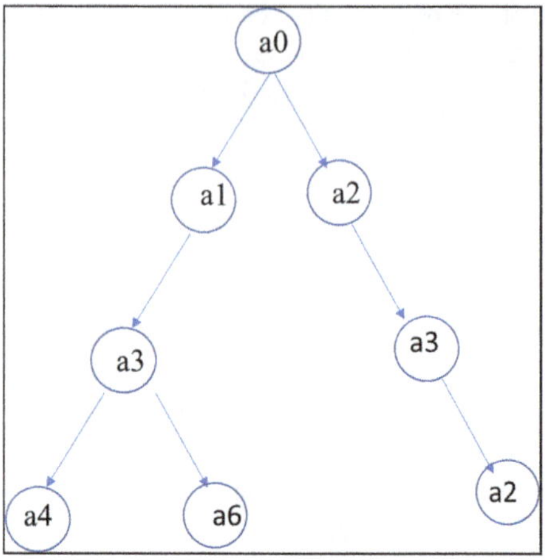

Fig. 17.1 Search tree example

$a_0 \rightarrow a_1 \rightarrow a_3 \rightarrow a_4$. The arrow denotes parent-child relationship among nodes. After processing the second sequence, node a_3 will have an additional child node a_6.

You can also select a sequence from the search tree by simply walking down the tree, selecting one node at a time. For the above example, start at the root node, a_0, and keep walking down the tree. Deciding to always take the rightmost branch, whenever faced with a choice, you would get the sequence (a_0, a_2, a_3, a_2).

For such a search tree, there is a unique path from the root of a tree to every node in it. You can see the unique path by walking upwards from a node to the root along the edges. So a node represents the corresponding complete path from the root. We will use nodes and paths (called *trajectories*) interchangeably, depending on the context and ease of explanation.

17.8.2 Monte Carlo Search Tree

Monte Carlo Search Tree (also referred to as *search tree* or simply *tree*) stores *trajectories* of states and actions only (i.e., it excludes rewards), starting from the current state in real environment. These are the *search tree trajectories* or simply *tree trajectories*. The *tree trajectories* are accumulated during simulation and always start with the current state in real environment.

An example of a trajectory in tree T would be Tr = [s_cur, a_0, s_1, a_1, s_2, a_2, …, s_N], where s_cur is the current state in real environment. s_N is the last state in the trajectory which may or may not be a *terminal state*. The trajectory Tr_S = [s_cur, a_0, s_1, a_1, s_2] in tree T is also a valid *tree trajectory*. Notice that Tr_S is a smaller trajectory that is fully contained in Tr. We say that Tr_S is a *sub-trajectory* of T_{au}.

For a *simulated episode*, a tree trajectory is first selected, and then the actual simulation starts from the last state in the trajectory and continues until the simulated

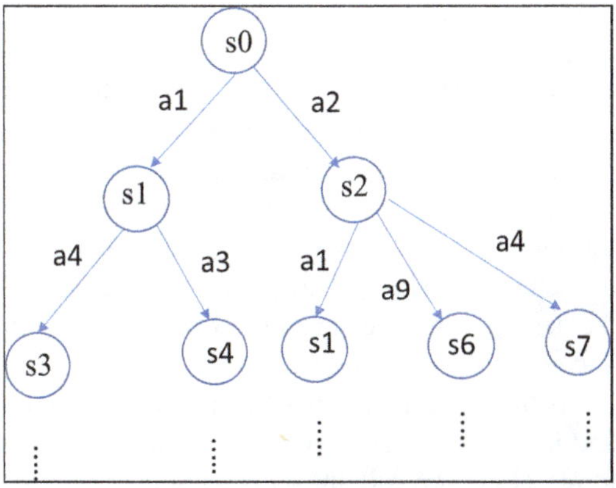

Fig. 17.2 Monte Carlo search tree (*Q*-values are not shown)

episode ends. You do not store entire *simulated episodes* in the tree. Instead you only store the *tree trajectories* that begin the simulated *episodes*. The trajectories are said to be *contained* in simulated episodes.

The tree grows (taller and wider) as you simulate more and more episodes, which in turn leads to improved *Q-value* estimates for all actions in the tree including the actions corresponding to current state in real environment.

In the search tree, nodes represent states and edges represent actions which are sometimes called *action edges*. Nodes and states are used interchangeably in this section. Figure 17.2 shows a Monte Carlo tree with states and actions. The *Q-value* of the action-state pair (a_1, s_1) is stored on the node for state s_1 (or you could store on action edge a_1). The root node of the tree always corresponds to the current state of the real environment. The size of the tree is roughly proportional to the number of simulated episodes.

The number of simulated episodes is only limited by the time available to Agent before providing the next action to the real environment. At the end of a round of simulation (or when time runs out), the Agent picks the best action A_S for the current state in real environment by identifying the root node's child with the highest *Q-value* and selecting the corresponding action.

Since the width of the tree is governed by the number of possible actions from a state, this approach is not feasible for video games with several hundreds or thousands of discrete actions because the width of the search tree will literally grow to billions of nodes even if the depth of the tree is in low single digits. However this approach works well for games such as *Go* and *Chess* because they only have tens of possible actions from most states.

17.8.2.1 Trajectory Values

The *tree trajectories* represent the most promising subset of possible trajectories that simulated episodes can start with. A good trajectory has higher *value*. The value can be *Q*-value defined as the average reward over all simulated episodes containing that trajectory. Suppose a sub-trajectory is contained in N_S simulated episodes, of which N_w episodes result in reward +1. Then the sub-*trajectory value* is N_w/N_S, the average reward. It also happens to be the probability of obtaining a + 1 reward in a simulated episode containing that sub-trajectory.

As the tree grows (explained in Sect. 17.8.3.2), it will have new tree trajectories. The probability of winning or losing (i.e., get a reward of +1 or −1) for such trajectories is not known. That is, these trajectories have *no value*. To assign *values* to such trajectories, you generate the simulated episodes starting from the last state of the new trajectory using *rollouts* (Sect. 17.7). Do several *rollouts* from the last state, and compute the value based on numbers of wins over the number of rollouts.

As the tree grows, a node n_1 that was once a leaf node becomes an *internal node*, and it no longer represents a new trajectory. We only do rollouts from new trajectories. However, rollouts will continue to be done on n_1's *descendent* leaf nodes (i.e., leaf nodes in the tree rooted at n_1). Any rollouts from n_1's *descendent* leaf nodes affect n_1's trajectory *value* since n_1's trajectory is contained in all the simulated episodes involving descendent leaf nodes.

The results (win/loss) of rollouts from a leaf node lead to updated trajectory *values* for its parent node, and all the way upwards, through a process called *Backup* explained in Sect. 17.8.2.2. Since the *value* of a node is same as the *value* of the corresponding tree trajectory, the trajectory value is stored on the node itself. Three *counts* are stored on a node, using which you can compute the value of the trajectory:

1. Number of episodes containing that node, i.e., the visit count
2. Number of winning episodes
3. Number of losing episodes

17.8.2.2 Backup Procedure

Consider a parent node and its children nodes. Clearly all trajectories containing a child node also contain the parent node. This lets you compute the *counts* for a parent node as a sum of corresponding *counts* of its child nodes.

Consider a parent node with just two children nodes c_1 and c_2, with winning counts win_c_1, win_c_2, loss counts $loss\_c_1$, $loss\_c_2$, and visit counts $visit\_c_1$, $visit\_c_2$. So, parent's win count is $win\_c_1 + win\_c_2$, the loss count is $loss\_c_1 + loss\_c_2$, and the visit count is $visit\_c_1 + visit\_c_2$.

If a child count was updated, say the winning count and total count both went up by 1, then you update its parent counts by simply incrementing the parent's winning count and visit count also by 1, and so on up to the root. This process of updating counts starting from a leaf node all the way to the root is called *backup or backpropagation*. Since newly created leaf nodes have no *value*, you can use rollout to assign values to such nodes, and after that you can perform *backup*.

17.8.3 MCTS Algorithm

You begin a round of simulations with a tree containing a single node representing the current state of real environment. The tree grows with every new simulated episode. Each simulated episode involves four phases – the *selection phase*, the *expansion phase*, the *rollout phase*, and the *backup phase*.

A simulated episode can be split into two trajectories: first trajectory is the *tree trajectory*, which starts with the current real environment state. The second trajectory is the *rollout trajectory* that starts at the last state of the tree trajectory. *Selection phase* generates the *tree trajectory*, and *rollout phase* generates the *rollout trajectory*. A simulated episode is started by selecting the most promising tree trajectory, one tree node at a time. This is known as *selection phase*.

17.8.3.1 Selection Phase (aka Tree Phase)

Tree root node *sCur* (current real environment state) becomes the first node in the chosen trajectory. Use the *tree policy* (described in Sect. 17.8.3.5) to pick the best action from among the actions available from *sCur*. Suppose you picked action a_0, and that resulted in next state s_1. You now have a trajectory: [*sCur*, a_0, s_1]. Now *walk* down the tree to node s_1, and repeat the process of selecting a new action, and therefore a new (state) node, using the tree policy. At each iteration add a new state-action pair to the chosen tree trajectory.

The selection phase eventually ends at a node, say s_N, when the tree policy can no longer be applied due to one of two reasons:

1. s_N is a leaf node.
2. One or more child nodes of s_N have no *Q-value*. The tree policy cannot be applied because it requires all actions (or, resulting states) to have some *Q*-values in order to pick the best action among them.

At this point you have chosen the tree trajectory portion of the simulated episode, and you enter the expansion phase.

17.8.3.2 Expansion Phase

If s_N is a leaf node, it may or may not be a terminal state. If it's a terminal state, the episode ends, and it already has the correct value – the reward. You can skip the next phase (*rollout phase*) and go straight to *backup phase*. If s_N is not a terminal state, *expand* s_N by adding child nodes to it, one for every possible action from $\mathrm{s_N}$. The new child nodes are leaf nodes, and they have no *Q-values*.

You now have a node, with children having no *Q*-value, either because you created these children just now or because the selection phase had ended due to reaching child nodes with no *Q-value*. Pick one of the children with no *Q-value*, say *sNew*, as the next node in the new trajectory, and move on to the *rollout phase*.

17.8.3.3 Rollout Phase

The purpose of a rollout is to give *some* *Q*-value for a tree node with no prior value. In this phase start with *sNew* that has no value. Now, identify an action a_0, using a random simulation policy (i.e., a policy that picks actions uniformly at random). Say, this action results in new state s_1. Now, apply the random simulation policy for this state, to get another action, and another state. Continue this way, until you reach a terminal state. This phase generates the *rollout trajectory*: *sNew*, a_0, s_1, a_1, s_2, a_2, …., *terminal_state*. Concatenating (i.e., joining) the tree trajectory (from *selection phase*) and rollout trajectory, in that order, gives you the simulated episode. Record the reward r after reaching the terminal state, and associate it with the node *sNew* from which you started the rollout. You don't have to store any of the states and actions from the rollout phase in the tree.

In this phase you can optionally do several rollouts instead of one to get a more accurate initial *Q*-value.

17.8.3.4 Backup Phase (aka Back Propagation Phase)

In this phase you start with the reward issued in the rollout phase. The rollout was done from node *sNew*. Use the backup procedure (described in Sect. 17.8.2.2) to update the counts at each node in the trajectory going backward up the tree from node *sNew*. Also update the *Q*-value of the node – the win probability – which is the ratio of win count and the visit count. When the *backup phase* is done, *sNew* will have a *Q-value* 0/1 for a losing rollout and 1/1 for a winning rollout. *sNew*'s *Q-value* will becomes more and more accurate as it gets updated in the *backup phase* of later episodes when it is no longer a leaf node.

17.8.3.5 Tree Policy

Given a tree node (i.e., state), the tree policy selects the best action from among the set of actions available (as edges) from the node. The selection is based on the *Q*-values of actions and an *exploration* factor to encourage selecting newer tree trajectories. A simple tree policy is the ϵ-greedy *policy* derived from the greedy deterministic policy. A greedy deterministic policy is defined as selecting the action with maximum win probability.

A more popular tree policy for MCTS is the *UCB1* policy. This policy is based on a function described by Eq. 17.11. The tree policy picks the action corresponding to child node with the maximum *UCB*1 *value*:

$$UCB1 = p(s_i) + C\sqrt{\frac{2\ln(N)}{n_i}} \tag{17.11}$$

where:

s_i denotes the child node
$p(s_i)$ denotes the winning probability of trajectories containing node s_i
N is the visit count for s_i's parent node
n_i denotes the visit count of node s_i
C is a constant

The formula encourages exploration by ensuring that *UCB*1 *values are* high for children with low visit counts. As visit count of a node gets higher, the chances of it being selected dwindle unless its probability of winning gets higher.

17.8.4 Pseudo Code for MCTS Algorithm

The pseudo code for a round of simulations using MCTS algorithm is given below. This pseudo code makes function calls to *expand the tree*, *do rollouts*, and *do backups*. The pseudo code for these functions is shown immediately following the pseudo code for the main MCTS algorithm.

```
T = Tree with single node, labelled with current real environment state
while (there is time for more episodes)
    current_node = root node of T  // start new episode

    while (current_node is not a leaf node, and all its child nodes have visit_count > 0) // selection
phase
        a_sel = TreePolicy(current_node) // pick action from current node
        current_node = child node corresponding to a_sel.
    end while

    // at this point above while condition is false so either current_node is a leaf
    //  node, or some of its child nodes have visit_count of 0

      if (current_node is a terminal state)
          Call Backup(current_node)

      else if (current_node is not a terminal state)
          if (current_node is a leaf node) //expansion phase
            Call ExpandTree(current_node) // creates new child nodes
     end if

     // Now, you have children with no value
        current_node = pick any child of current_node with visit_count of 0
```

```
        Call Rollout(current_node) //rollout phase

           Call Backup(current_node) //backup phase

        end if // end current node is not a terminal state

    end while // more episodes

End of main algorithm

  function TreePolicy(input  current_node)
          Let A be the set of actions from current_node.
          foreach action a_i in A
              get the UCB1 value for a_i
      end foreach
        Choose the action corresponding to maximum UCB1 value

  End of tree policy function

  function Rollout (input node)
    cur_state = input node
    while (cur_state is not a terminal state)
              Pick a random action a, and apply to environment.
              cur_state = new environment state due to action a.
         end while

       // cur_state is a terminal state. You have a reward value
       If reward is +1, increment input node's win count.
       If reward is -1, increment input node's loss count.
       Increment input node's visit count

end function

End of rollout function

function Backup (terminal node, reward)
    n = parent of terminal node
    while (n != root)
         increment n's visit count.
         if (reward is +1) increment n's win count
         if (reward is -1) increment n's loss count
         n = parent of n
  end While
end function

End of Backup Function.
```

17.8.5 Parallel MCTS Algorithms

MCTS algorithm is inherently sequential. It has to go through the four phases in sequence. However, there are ways to achieve parallelism for faster computations. Parallelism is a key factor in the popularity of MCTS algorithm.

Root parallelization involves creating N copies of the search tree and running the sequential MCTS algorithm on all of them in parallel. Then, merge the results from parallel runs onto one tree.

In *thread parallelization* a single search tree is traversed using multiple execution *threads* (such as *operating system threads*) in parallel. Since multiple threads cannot modify a node at the same time, node locking has to be implemented. That is, a thread traversing the tree will lock down all nodes as it traverses the tree. The thread unlocks all the nodes once it has done the node updates during *backup* phase. A major downside of this approach is possible slowdown due to *thread contention*, i.e., if a thread wants to update a locked node, it has to wait for the node to be unlocked by the thread that locked it. Another downside is that many threads may traverse the same trajectory, bringing no additional value and wasted compute time.

A popular variant of *thread parallelization* is called *MCTS with virtual loss*. In this version, whenever a thread locks a node, it temporarily reduces the node's *virtual* value. Essentially, it tags the node as a low-valued node so that other threads stay away from the node. This reduces contention. The thread removes the virtual value whenever it unlocks the node. Another advantage is that *exploration* is encouraged, as other threads will attempt to select different trajectories, as a result of avoiding the locked nodes due to low virtual values.

17.9 MCTS Tree Values for Two-Player Games

Consider any two-player game such as chess where actions are the moves made by either players, and the environment state is the board configuration. The environment state should also include whose turn it is next. So add an additional state variable (a Boolean value) to the environment state vector, to represent whose turn it is to play next. Since you must simulate the moves of both players, store the actions of both players in the same tree.

Suppose root state of the tree s_0 indicates that it is white's turn to play. That is, in the real environment, black made a move, which resulted in this state. A tree trajectory would then be $[s_0, a_0w, s_1, a_0b, s_2, a_1w, s_3, a_1b, \ldots]$, where the moves alternate between the two players (a_0w represents move a_0 made by white). The last state in the trajectory represents a win, a loss, or a draw for each of the two players.

The Q-value of a node will be different for the two players. The win probability of one player equals the loss probability of the second player. Note that the win probabilities of both players may not add up to 1, as some simulated games can end in a draw. Similarly, the UCB1 values would also be different for both the players.

The tree policy is supposed to identify the best trajectory by selecting nodes with maximum *UCB*1 *value*. In a two-player context, all nodes also contain information about whose turn it is. From any given node, the tree policy should pick the child that has maximum *UCB*1 value for the player whose turn it is now.

The backup procedure updates the win count individually for each player, at each node, depending on which player won the rollout.

17.10 Alpha Zero

Google *DeepMind* developed a series of computer programs over a few years that learnt to play the games of *Go* and *Chess* at a superhuman level with little training from humans. These programs incorporate reinforcement learning techniques combined with deep convolutional neural networks to approximate value function and policy network. The main programs in chronological order are *AlphaGo*, *AlphaGo Zero*, and *Alpha Zero*.

Each newer version was a much stronger player than previous version, was simpler, and used lesser human input (for training examples and hand-tuning of features and hyperparameters for a specific game). *Alpha Zero*, the latest (at the time of writing this book) and the simplest, is generic enough to play *Go*, *Chess*, and *Shogi* without any game-specific hyperparameter tuning.

Key ideas behind *Alpha Zero*'s algorithm are explained here, without going into implementation details. *Chess* is used as an example for this section. Refer to Google DeepMind's papers for more details.

In two-player game terminology, the action taken by one player (e.g., move a piece on the board) is called a *ply*. The word *move* in chess denotes two plies made in succession (one by each player). In this section we will use the term *move* to mean a *ply* for clarity.

17.10.1 Overview

Alpha Zero also relies on *MCTS*, though with a few differences. *Alpha Zero* makes use of deep neural networks, has a different *MCTS Tree* policy, and does not do rollouts from leaf nodes (instead it evaluates the leaf nodes using a value function computed by neural network).

17.10.1.1 Value Function and Policy Network

The following two functions are created using a single deep neural network that takes in the board state as input:

1. The value function to estimate the probability of a specific player winning from the current board state.
2. The policy network, to estimate the probability distribution of actions possible from a given board state. The actions here are the legal moves.

A value function is used in addition to policy network because game play purely based on either of the two functions does not lead to strongest possible program. It is advantageous to have both the functions and combine them in a particular way (explained in Sect. 17.10.1.4) to make a stronger program.

Section 17.10.1.3 explains the generation of labelled examples for both functions, for training the neural network.

17.10.1.2 MCTS Search

MCTS search relies on both the value and policy functions to expand the tree and pick the next node in the tree trajectory. Policy function from neural network is used to select the nodes in the trajectory. The values at leaf nodes are obtained by evaluating the neural network (value function), instead of via rollouts, which greatly improves the accuracy of leaf node values because the values are effectively equivalent to performing a large number of rollouts from that node. The *backup* is done as usual based on those values. A key assumption is that moves chosen by MCTS search will be stronger than using just the value function or the policy network.

17.10.1.3 Self-Play and Training Data

MCTS search depends on precomputed value and policy functions implemented by the neural network. The supervised training examples for the neural network are generated using *self-play*, i.e., the program playing with itself. Each move in a self-play game generates a training example for both policy network and value function. A deep neural network generally requires millions of training examples. Since each chess game lasts anywhere from 50 to 200 moves, you would need 100K games of self-play to generate about 10 million training samples.

17.10.1.4 Iterative Improvement Loop

Start with a neural network N_0 with random weights, which should result in random estimates for value function and policy network. Now suppose you execute a set of *self-play* games using MCTS search to select each move (relying on the value function and policy network based on N_0) to generate a set of training data TR_1. The moves in TR_1 should be a little better than random moves. Now, if you train the existing neural network further using TR_1, then you would get a slightly better neural network N_1. If you use N_1 in self-play games to generate training set TR_2, you expect TR_2 to have better moves than TR_1. If TR_2 is used to further train the neural network, you should get an even better neural network N_2. Continuing this iterative process you should get better and better neural networks.

In a real game, the moves are chosen with MCTS which further boosts the strength of the moves as compared to the moves from using the value function or the policy network.

17.10.2 Aspects of Alpha Zero

Alpha Zero uses a 40-layer deep *CNN* based on *ResNet*, which generates two sets of outputs:

1. A single scalar for the value network ranging from -1 to $+1$, representing the expected reward from the state. This value reflects the probabilities of a win, loss, or draw (e.g., a positive value closer to 1 indicates higher chance of winning).
2. A probability distribution of next moves (policy network) that comes out of a *softmax* layer. It outputs a fixed number M of probability values, where M is the number of possible moves from any board state. Thus it represents the superset of all possible moves, many of which may be illegal for the current state. The probabilities for illegal moves may be small if not zero as a result of training. Therefore the distribution of the M values is post-processed to remove the illegal moves, and the remaining legal moves are renormalized to form a probability distribution.

17.10.2.1 Supervised Training

You've learnt earlier that moves in self-play games become the training examples. The neural network is essentially two networks in one – it includes both policy network and value function, and they both have a common input (the board state). You create a single training example for every move that trains both networks.

Suppose the board state is S during self-play game. The input in the training example is S. The expected output of training example has two parts, one for policy network and one for value function.

The expected output for policy network is simply the move m made from state S. The expected output for value network, r, depends on who's turn it is to play next (captured as part of state), i.e., who made the move m. If the player who made the move eventually wins the game, then the r is $+1$. On the other hand, if this player lost the game, then r is -1. If it's a draw, then r is 0.

So, each move creates a combined training example with expected output as the pair (m, r).

A strategy similar to *experience replay* (see Sect. 17.5) is used to identify mini-batches for training (through stochastic gradient optimization) to avoid the problem with successive moves in a self-play game being correlated.

17.10.2.2 Loss Function

Since there is only one neural network, there is a single loss function that is a linear sum of the loss functions for value function and policy network.

The objective is to minimize the prediction errors made by the neural network.

The value function output is a scalar value, ranging from -1 to $+1$. Its loss function is simply mean squared error. The policy network output is a probability distribution, and therefore its loss function is cross-entropy. The final loss function is given by Eq. 17.12. The first term, $(z - v)^2$, is contribution of error in value function. The second term, $-\pi^T \log p$, represents the *cross-entropy*, and the last term is the *regularization* term to avoid overfitting:

$$l = (z - v)^2 - \pi^T \log p + c||\theta||^2 \tag{17.12}$$

where

z is the value function from self-play (+1, 0 or -1).
v is the predicted value function output(ranging from -1 to +1).
π is the probability distribution of moves from self-play.
p is the predicted policy network output.
c is a constant.

17.10.3 MCTS Search

The Alpha Zero version of MCTS also has four phases (selection, expansion, rollout, and backup: explained in Sect. 17.8.3), but differs from conventional MCTS search in a few aspects.

17.10.3.1 Node Value

In conventional MCTS you usually store three values, visit count, win count, loss count, and they are used to compute the Q-value for each player. Alpha Zero stores the visit count N and the cumulative sum of values W (explained in Sect. 17.10.3.5). These are used to compute the Q-value as W/N. Also, *Alpha Zero* stores the values on edges instead of nodes, though this difference does not really matter.

17.10.3.2 Selection Phase

There are two tree policies in effect – one for selecting the next move in the trajectory (call this *trajectory policy*) and another for selecting the best move for the real environment (call this *root policy*).

The *trajectory policy* of a node with state-action pair (s, a) depends on the following values:

1. Visit count N
2. Probability $p(s,a)$ obtained by evaluating the neural network
3. Sum of visit counts ($N\text{Sum}_C$) of the node's children nodes
4. Sum of values W

Now compute two quantities based on above values:

1. *Q-value* $Q(s,a)$: This is the *exploitation* term, obtained as W/N.
2. *U-value* $U(s,a)$: This is the *exploration* term to help MCTS search explore new trajectories and is given by Eq. 17.13.

$$c_{\text{PUCT}} * p(s,a) * \frac{\sqrt{N\text{Sum}_C}}{1+N} \tag{17.13}$$

where:

c_{PUCT} is a constant to control the preference for exploration.
The policy picks the action that maximizes the sum $Q(s,a) + U(s,a)$.

Root policy is used for picking the best child node of root node, to be passed on to the real environment. This policy only depends on visit counts of root node's children and a *temperature* hyperparameter τ. The temperature parameter is only used in self-play and is set to either 0 or 1. When τ is 1, the child nodes are chosen stochastically in proportion to visit counts. If τ is 0, the node with maximum count is always chosen. τ is set to 1 for the first 10s of moves to encourage exploration and set to 0 for the rest of the self-play game so that it can pick the best move. All other values on the nodes such as *Q-values* are completely ignored in this policy since their effect translates to visit counts on nodes. That is, nodes with higher values will naturally accumulate higher visit counts which then lead to higher visit counts on nodes closer to the root including root's children.

Note that in real play (i.e., when playing against an opponent), the node with maximum count is always chosen.

17.10.3.3 Expansion Phase

This phase is same as in conventional MCTS. The leaf node is expanded as usual.

17.10.3.4 Evaluation Phase (Replaces the Rollout Phase)

Instead of rollouts from a new leaf node to estimate its initial Q-value, obtain its value function by evaluating the neural network, and use that as the initial Q-value (to be used in backup phase).

17.10.3.5 Backup Phase

The visit counts and W values of all nodes visited in this phase are updated. The visit count of a node is incremented by 1, and the W value of a node is updated by adding the W value of the leaf node. The Q-value of a node is recomputed, as both numerator and denominator of W/N have been updated.

17.10.3.6 Parallel Execution

Given the large amounts of training and self-play games needed, *DeepMind* parallelized all the main tasks, including MCTS search, the self-play games, the neural network evaluation, and neural network weights updates so as to finish the training in a reasonable amount of time (hours or days instead of months or years!). This parallelization used *MCTS with virtual loss*.

Reinforcement learning has made possible some of the capabilities that were thought to require human intelligence, such as computers' ability to play games and autonomous driving. This has dramatically increased interest in RL. RL is a relatively complex topic, utilizing several other ML concepts and ideas, read in earlier chapters. So, it is normal if the chapter seems overwhelming in the first read. RL continues to be an area of very active research, and this chapter has only covered a small fraction of RL. New RL algorithms continue to be published as research progresses. Hopefully this chapter provided you with a good conceptual understanding of reinforcement learning.

Chapter 18
Designing a Machine Learning System

In the previous chapters, you have seen various algorithms and how they apply to specific problem domains. This chapter will help you get into the finer details of designing a machine learning system. The concepts explained in this chapter are less about individual algorithms; they are about making choices for implementing your algorithms.

18.1 Pipeline Systems

The prior chapters have explained several approaches toward solving individual problems in machine learning. In most real-life problems, you need to break down the problem into various components and apply a combination of techniques. These techniques are applied in sequence that form a pipeline where output of one algorithm stage becomes input to the next stage in the pipeline. This chapter introduces you to a simple method to identify stages with performance issues in the pipeline to improve your results of the whole system.

Consider a system that you are designing to identify the number plate of a speeding vehicle. For this, you will have to design four components:

- Speed detection: This component will monitor the speed of the oncoming/passing vehicles and will trigger whenever the speed goes above the prescribed speed limit. Depending on the policy, instead of just a speeding/not speeding decision, it might record the actual speed. And, for speeding vehicles, it will capture the image of the car.
- Number plate segmentation: This component will extract out the portion of the camera image that has the number plate. The rest of the car image is not of much importance.
- Identifying the vehicle number: This component will recognize the characters on the number plate – in order to identify the registration number of the vehicle.

G. Rebala et al., *An Introduction to Machine Learning*,
https://doi.org/10.1007/978-3-030-15729-6_18

- Generating the ticket: This component will access the transport department's records to generate the ticket for the vehicle and post it to the right contact address.

These components form a pipeline, in the sense that the output of the first component feeds into the second component and so on.

As you implement and evaluate your system, you find that the whole system provides an accuracy of only 68%. Obviously, you want to increase the overall accuracy. The question is, which component of the system should you improve first.

18.1.1 Ceiling Analysis

Just like any other analysis involving performance tuning, you need to identify a single parameter (or figure of merit) that you will use to evaluate your system. In the current example, we are considering being able to accurately fetch the registration details for speeding vehicles as the parameter of interest.

You assume that the first component (speed detection) works with 100% accuracy and manually feeds the *correct* data to the second component. This *correct* data may have been manually obtained. Now see how the accuracy of the whole system improves. Say, the accuracy now improves to 72%. This says that by improving the first component, you will gain a maximum improvement of 4%. This 4% represents the ceiling of the improvement that you can get by improving the first component.

Now, assume that the second component also generates output with 100% accuracy (i.e., both first and second component are working with 100% accuracy). And, you can once again manually generate the *correct* output from the second stage and feed it into the next stage. Let us say that the overall accuracy of the whole system now rises to 80%. This means that by improving the second component (or stage), you can get a maximum improvement of 8% from the whole system. This 8% once again represents the ceiling of improvement achieved through improving second stage.

You use the same technique to identify the ceiling of improvements possible for each stage. And, now, you should work on improving the stage which has the highest ceiling value. Improving this component will provide you the highest gain in the overall accuracy.

18.2 Data Quality

In the world of machine learning, the richness of training data is much more important than minor improvements to algorithms – for better prediction. Data collected in the real world is often dirty with *duplicate data*, *missing data*, *expired data*, and *inaccurate data* affecting data consistency, resulting in poor outcomes for the algorithms.

Data models used in business intelligence are dependent on integrity models of relational databases with rules based on functional dependencies of the data. For example, data integrity rules specify requirements of foreign key in schema specification like sale of item should have an item information associated with it. Functional rules specify that if a merchandise is returned, then the date of return should be later than the date of sale.

Significant work is done in sanitizing data in business applications with satisfactory techniques in replacing null values, identifying missing data, and identifying duplicate data. These are techniques based on rules created by domain experts based on their understanding of data dependencies, consistency requirements, and functional requirements. Rule sets are created based on the requirements of the domain. As the number of data sources increases, the rule sets get complex and are hard to create as these rule sets can be extensive.

The rules for data cleaning are also dependent on each other and need to work in tandem. For example, finding duplicate data will work better if the missing data is replenished. On the other hand, missing data is better replenished if the duplicates are identified first. It often helps if the rules are executed together to improve the quality of the data. Often, the execution order in the rule execution makes the difference in resultant data quality. Machine learning techniques like *Bayesian models* can be used to apply multiple rules together in improving data quality.

Missing data or nulls in structured data can be filled in with a known median for that variable or other statistical parameter. It is possible to decide to remove the data with nulls depending on the domain or application. For example, in a time series or a sequence, missing values can be duplicate of previous value or average of values adjacent to missing data. In a non-sequential data like user preference in shopping, the data point with missing or uncertain user ID may be deleted from dataset.

Duplicate data is detected based on the domain. In a simple case, if the user ID matches in the database and there is expected to be only one entry based on user ID, then more than one entry will be duplicate entry. Depending on the domain model, the data can be merged or one of the entries will be deleted.

Data profiling and cleaning are performed based on the domain knowledge. For example, user name John Doe can be written as Doe John or J.D. Using other information available in the dataset, the user name can be identified as the same user or a different user. Complex rules based on the domain information will be used in resolving the conflicts or profiling the data.

Anomalous or incorrect data either because of errors in entry or incorrect labelling of data also results in similar errors in modeling. Anomaly detection techniques discussed in Chap. 13 can be used to remove anomalous data. Domain rule-based techniques can also be used to remove the anomalous data points.

18.2.1 Unstructured Data

Machine learning algorithms are increasingly being used on unstructured data scraped from the Internet. This data could be coming from sites like Twitter and Facebook which is completely unstructured. Data source could be websites tracking data based on cookies that may or may not identify a specific user and is also unstructured. Data could be from sources tracking information like retail stores that use discount store cards to track spending patterns that is associated with a discount card. Data could be from credit rating agencies that track reported data on individuals or companies from different credit sources.

This unstructured data, also termed as *Big Data*, requires additional techniques to clean the data to make it useful for machine learning algorithms. The data quality and precision requirements change based on the application. For example, application identifying fashion trends to predict most efficient store display needs only generic trends based on past customer profiles and pace of change in tastes. Whereas identifying credit terms to extend a customer will need a more precise financial profile of the customer in question along with the trends of defaults within the class of profile.

18.2.2 Getting Data

Performance of algorithms depends on quality of training data. Hence, it is very important to spend some time in thinking of ways to collect a lot of labelled data. For character and image type of problems, while it is very easy to get a lot of data, the process of labelling them could be very costly, if humans have to manually label each of these data points.

As machine learning is becoming popular, there are various websites, which provide open-source labelled data – especially for pictures/images and text characters. Feel free to make use of these sources.

Often, you can create many more instances of data from a set of already available data. Example, for characters, you can create a large set of labelled training data through:

- Use many different fonts for the character.
- Put random background for these characters.
- Put slight rotation or blur some portions of the character.
- Apply some distortion, e.g., extend the character only vertically etc.

So, for a character *A* – by taking 10 different fonts, and putting 3 different types of background, you got 30 samples of *A*. As you can imagine, with little creativity, you can easily get ten times the original dataset, obviously assuming that the original dataset was itself unique, rather than derived from each other.

You can also use a crowdsourcing technique such as Amazon Mechanical Turk or Figure Eight (formerly called CrowdFlower) to get a huge quantity of labelled data. Or, you could collect and label your own data – which could be very time consuming. Best is to be able to tap into already existing dataset (if available) that you can repurpose.

Also, make sure that the data that you are collecting for training is a fair representation of the sample space on which you finally intend to apply your algorithm. For example, you want to create an application for facial recognition that is supposed to work across the globe. You want to ensure that it is trained on a large dataset representing a wide variety of demographic variation across ethnicity, age, gender, dressing styles, etc. There have been embarrassing cases of face recognition machines not working well on people of specific ethnic background!

18.3 Improvisations over Gradient Descent

In Chap. 3, you saw the concept of slope and how slope is used to refine the value of parameters, till we reach close to the bottom of the cost curve. This section will show you some improvisation for reaching the cost curve faster. Section 4.3 had also shown some improvisations. Those improvisations were more in the form of how much data to consider for computing error, so that each iteration is faster. This section, though, is about how to get slightly better corrections during each iteration, so that you need lesser number of iterations to reach the desired values.

The concepts explained in this section are more meaningful if you are using Stochastic Gradient Descent, as explained in Chap. 4. Recall that Stochastic Gradient Descent works on one data-point at a time, and hence, may not take the steepest path toward solution, rather it takes a meandering path. It is also possible that Stochastic Gradient Descent could oscillate – once it reaches close to the solution, however, it might not really reach the actual solution. The improvisations mentioned here will help jump over such local minima – in addition to reducing the meandering of the solution.

18.3.1 Momentum

Conceptually, when using Gradient Descent, you are considering the slope at your current point and then use that slope to go down the cost function curve. Think of it like a ball on a hill, and then, the ball will roll down the hill in the direction, where the slope is the highest. Extending this analogy, if the ball was already coming down the hill, at any given point, its speed and direction will depend not just on the slope at this point but also on the speed and direction gathered so far. This is what is meant by momentum (though, strictly in physics, momentum considers the mass and velocity of an object).

The idea thus is, while determining the correction, consider the current slope as well as the previous correction (indicating the speed and direction with which the ball reached here).

Equation 18.1 shows the correction as in regular Gradient Descent, and Eq. 18.2 shows the correction after considering the momentum aspects. Equation 18.3 shows the updated value of the parameters after applying the correction. You will notice that Eqs. 18.1 and 18.3 together form Eq. 3.10 (your original concept of Gradient Descent based correction):

$$\text{Correction} = \alpha \frac{\delta J}{\delta \theta_j} \tag{18.1}$$

$$\text{Correction with momentum} = \gamma * \text{previous correction} + \alpha \frac{\delta J}{\delta \theta_j} \tag{18.2}$$

$$\theta_j = \theta_j - \text{correction} \tag{18.3}$$

Equation 18.2 introduced yet another hyperparameter, γ, which represents how much of the previous correction should influence the current correction. Its value can be of the order of 0.9.

Effectively, the concept of momentum aids in reaching the solution faster because:

- If the slope at the current position is in the same direction as the previous iteration, you take bigger steps (compared to what you would have taken, if you considered only the slope).
- And if the slope at the current position changes direction, then your step size is reduced.

Some people also recommend starting with a smaller (say 0.5) value of γ and then gradually increasing it up to 0.9. This is because, at the beginning, you might already be taking larger steps (due to higher slope – away from the minima).

18.3.2 RMSProp

RMSProp is an algorithm to update the learning rate α iteratively or adaptively. The equation for updating θ as given in Eq. 3.10 is modified to the one given in Eq. 18.4:

$$\theta_j = \theta_j - \frac{\alpha}{\text{RMS of prior gradients} + \epsilon} \cdot \frac{\delta J}{\delta \theta_j} \tag{18.4}$$

where

RMS stands for root mean square
ϵ is a very small value to prevent the possibility of division by 0

As you can see, the learning rate itself is being iteratively adjusted during each iteration. Instead of computing the RMS of all gradients since the start of the algorithm, the algorithm suggests to have an approximation given by Eq. 18.5:

$$(\text{RMS}_{\text{current}})^2 = \gamma * (\text{RMS}_{\text{previous}})^2 + (1-\gamma)\left(\frac{\delta J}{\delta \theta_j}\right)^2 \tag{18.5}$$

The γ in Eq. 18.5 serves a similar purpose as in momentum, except that instead of impacting the slope contribution, it is impacting the learning rate contribution. The learning rate is adjusted by a contribution of both prior gradients and current gradient, while causing an exponential decay in the contribution of older gradient values.

You have learnt about adaptively adjusting the learning rate during each iteration. It is also common to have a unique learning rate for each parameter. If you decide to have unique learning rate for each parameter, while computing the RMS, you should consider only the gradients with respect to the corresponding parameter, rather than the overall gradient.

18.3.3 ADAM (Adaptive Moment Estimation)

Before going into the mathematics of this algorithm, let us try to explain it conceptually. Equation 3.10 had given an equation for iteratively refining the values of the parameters and is rewritten here:

$$\theta_j = \theta_j - \alpha \cdot \frac{\delta J}{\delta \theta_j} \tag{3.10}$$

In this algorithm, instead of using the current slope, you use the average of prior gradients, and instead of using a fixed learning rate, you use an adaptive learning rate, as explained in Sect. 18.3.2.

So, the equation for finding the next updated values for the parameters will be given by Eq. 18.6:

$$\theta_j = \theta_j - \frac{\alpha}{\text{RMS of prior gradients} + \epsilon} \cdot m \tag{18.6}$$

where

m represents the average of slope
RMS is root mean square (explained in previous section)

Once again, instead of retaining the average of slopes across all iterations, you can compute a running average, using a decayed contribution of older values, through Eq. 18.7:

$$m_{\text{current}} = \beta_1 * m_{\text{previous}} + (1 - \beta_1) * \frac{\delta J}{\delta \theta_j} \tag{18.7}$$

Note that you can have different values for β_1 and γ *(used in eqn 18.5)*, though both of these are expected to have values closer to (but less than 1). In many literatures, β_2 is used instead of γ, in order to distinguish this hyperparameter from other hyperparameter (say: the one used in momentum).

For the sake of completeness, the word "moment" in the name of this algorithm comes in because mean and squared mean are mathematically called first and second moments. In this case, instead of the actual moments, you are taking an estimate of these moments.

18.4 Software Stacks

With the knowledge that you have by now, you can implement your own machine learning algorithms, using any of the software tools, package, and language that you are most comfortable with. However, there are some software frameworks that already exist – just for machine learning algorithms and applications. You can make yourself familiar with one of these popular frameworks which should increase your productivity.

Most of these frameworks also make optimal use of hardware mentioned in Sect. 18.5 like FPGA, TPU, and GPU to take advantage of the parallelism available in the hardware resources. Additionally, machine learning algorithms are enhanced to run in parallel in each stage of the pipeline to improve processing time. For example, when processing one terabyte of data, if the algorithm can break up the data in 100 chunks of 10 gigabytes and combine the results without loss of accuracy, then the algorithm can take advantage of massively parallel processing capabilities of hardware.

TensorFlow, *MxNet*, *Theanos*, *CNTK*, *Caffe*, *pyTorch*, and *Chainer* are some of the frameworks that support constructs for parallelization of algorithms.

18.4.1 TensorFlow

TensorFlow is a framework developed by Google Brain team that takes advantage of GPUs and TPUs natively. TensorFlow is built using Python to conduct deep learning research on image recognition and language translation and can be used for general purpose modeling for deep neural networks. TensorFlow gained considerable following in machine learning community because of its efficient processing in Google data centers.

TensorFlow provides a framework to define computational graph with nodes of graph representing computational units and edges representing tensor flows between the nodes. A tensor is an *n*-dimensional array with 0-d, representing a scalar value; 1-d, representing a vector; a 2-d, representing a two-dimensional matrix; etc. *TensorFlow* achieves computational parallelism by constructing graphs of execution with each of the nodes as computational units and placing the computational units in a GPU or TPU for execution. Since *neural network* models are represented as directed graphs anyway, the mapping of the computational units to neural network nodes is straightforward. This achieves great speedups in the processing for speech recognition, image processing, etc.

TensorFlow lags behind other frameworks in benchmark testing due to its Python abstractions that do not map to TPU executions very cleanly. Language bindings are also not as extensive as some of the other frameworks like *MXNet*. Since *TensorFlow* executions are parallelized and executed by framework, and not imperative programming flow, debugging can be difficult. TensorFlow is extensively used by research community, and its features, like *TensorBoard* for visualization, checkpoints for managing experiments/troubleshooting, are of great value for model development.

18.4.2 MXNet

MXNet is an Apache Software Foundation project supported by Amazon and currently provided as part of Amazon Web Services offerings. This project also has support of Microsoft and is offered in the Azure cloud.

MXNet supports both imperative and symbolic programming. Symbolic programming allows nodes to be defined by computational functions and executes a graph with nodes connected by binding them to values through abstract graph structures. These are very useful in defining and representing deep neural network structures. Imperative programming on the other hand is most familiar with software engineers with computational flow defined through imperative statements. The *NDArray* API of MXNet supports the traditional programming model of type $C = A + B$ where A, B, C represent data structures like arrays.

MXNet offers libraries that help developers to take advantage of CPUs and placement for parallelism and computational efficiency. *MXNet* supports multi-GPU training, placement of data structures, and automatic differentiation for diagnostics. *MXNet* has performed better on the benchmark tests because of its ability to take advantage of GPUs better.

MXNet also offers large set of language bindings for developers to work with their favorite language. *Gluon API* has simplified invoking machine learning functions for common applications like recommender systems. Framework also makes it easy to program for CPUs and then convert the program to take advantage of GPUs for performance.

MXNet has not received the community support as much as *TensorFlow*, and it is also not designed for TPUs as special purpose acceleration. This could change in the future with support from *Amazon* and community support for special purpose processing units.

18.4.3 pyTorch

pyTorch is a framework developed by Facebook for deep learning applications like image recognition, speech processing, and language translations. *pyTorch* is written in Python supporting dynamic graphs which means that it can build a new graph on forward pass for the neural network. *pyTorch* supports Python's object-oriented functionality, to extend a feature, simply define a class, and extend *pyTorch*.

pyTorch is built on top of *Torch* and *Caffe2* image processing library built by Berkley and enhanced by Facebook and community. pyTorch model is based on tensors and with imperative model that supports dynamic graph generation to add or remove hidden layers during training to improve accuracy. The imperative model has made it easy to start with the model and debug and had gained traction with developers in modeling deep neural networks.

pyTorch also supports automatic gradient computation similar to *TensorFlow* or *MXNet*. It has strong GPU integration with the libraries for computation. *pyTorch* has built-in support for Stochastic Gradient Descent variations like RMSProp, Adam, momentum, and adaptive learning rates.

pyTorch has good community support but not as strong as *TensorFlow* and performance not as good as *MXNet* in large-scale environment. *pyTorch* is a good choice for developers with software development background for smaller learning curve and good development tool support.

18.4.4 The Microsoft Cognitive Toolkit

The *Microsoft Cognitive Toolkit*, used to be *CNTK*, is a framework from Microsoft that has support for processing for very large datasets with distributed in-memory processing engine like *Spark*. Microsoft Cognitive Toolkit has very good language bindings to Java, C++, C#, and Python. It supports integration with Microsoft IDE tools and runs on Azure cloud. It has support for GPUs and provides scalable performance.

The licensing terms are not open-source variants, and the community support is not extensive. It is not clear (to us) if the framework is being used extensively by other companies. Microsoft also supports other frameworks like *MXNet* on its cloud platform. Developers focused on Microsoft platforms and Azure cloud will find the framework a good choice for deep neural network model development.

18.4.5 Keras

Keras is a high level neural network API written in Python that abstracts some of the complexities of *TensorFlow*, the *Microsoft Cognitive Toolkit*, *MXNet*, or *Theano*. It is designed primarily for designing and rapid prototyping that can take advantage of CPU or GPU and support CNN as well as RNN deep neural network models.

Keras has become very popular because of its developer friendliness, modularity, and Python integrations. It is a good choice if the underlying framework is selected, and Keras provides the learning curve and adoption efficiencies. *Keras* is consistently slower than the frameworks like *pyTorch* but provides developer efficiency. It is possible to imagine libraries from *MXNet* like *Gluon API* will bridge the gap and provide the ease of use for the developers to gain efficiencies in both development time and execution.

18.5 Choice of Hardware

Most of the algorithms used in machine learning are very compute intensive, and they also operate on a large set of data. The large set of data could be the actual data itself or intermediate computations or weights for neural network kind of designs. For an efficient implementation of machine learning systems, you need to choose an appropriate hardware. The choice of your hardware system should consider both the compute efficiency, as well as data movement efficiency (refer to Sect. 1.7). Whichever hardware do you choose for your machine learning implementation, you would need a PCIe connection on the hardware – to be able to connect to a host machine which controls the processing and access to huge memory (to store a huge amount of data).

18.5.1 Traditional Computer Systems

Traditional computer system (PCs, laptops, etc.) is the first and the most obvious choice. It allows you to program your algorithm in your choice of languages (C, C++, Python, etc.) and execute on the processor.

The best advantage of such systems is that these are most commonly understood, used, and available. So, there is no need for a long learning curve. There is a high degree of flexibility for you to tweak your algorithm; you just have to edit your code, recompile, and execute. You can debug your algorithm using a plethora of debug tools available.

However, in spite of such obvious advantages, such computers are not deployed very widely in real-world machine learning problems. The main reason for lack of popularity is a high degree of parallelism that is offered by some of the other hardware systems mentioned below.

18.5.2 GPU

GPUs offer a very high degree of parallel compute capability. GPUs were originally designed for graphics (i.e., video) applications. Since they were dealing with pixels and graphics, they were designed for parallel processing. As machine learning became popular, many users saw an advantage of using GPUs for its inherent parallelism (over PCs) and started moving toward GPU.

GPU manufacturers, in turn, saw an opportunity in machine learning world and started providing support for such applications, in addition to the original intended graphics market.

The high degree of compute along with parallel processing can be utilized for many matrix operations. This should explain why many machine learning problems are expressed in matrix forms. Besides ease in expressing, they lend themselves very easily to such compute structures that have high degree of parallelism.

GPUs are made of many cores, each of which can perform a compute operation. This presence of many cores is what allows high compute power as well as parallel operations. Each of these cores can operate on its own set of operands. However, the operation across all the cores is still the same. So, while 100s of cores could be doing addition, they can add different set of operands. This mechanism is called *SIMD* (single instruction multiple data). SIMD capabilities of GPUs are generally useful, when you have a high degree of uniformity in your operations; matrix operations usually fit the bill.

GPUs also possess local memory. The advantage of having local memory is proximity to the compute unit, and hence, you won't waste time in fetching data from external/secondary memory or in writing into external memory.

GPUs also allow for floating point operation. Floating point allows you to represent a much wider range of data values (though at the cost of accuracy) for the same number of bits. Thus, specially, at the training stage, where the various weights/parameters could fluctuate over a wide range of data, floating point could be useful. There is some controversy with respect to that understanding, in the sense that there have been some studies which show that instead of worrying about being able to deal with a large swing of data, for some applications, you get more reliable results, if you use more accurate values.

Since GPUs have been addressing the machine learning market for a while, they have a reasonably good software ecosystem that provides good integration with some of the most popular packages used in machine learning. So, you could still write your code with your familiar package, and you can take the implementation to the GPU with reasonable ease.

18.5.3 FPGAs

FPGAs are another class of programmable devices, which are getting popular in the world of machine learning. Like GPUs, they also provide very high degree of compute power, including parallelism. The on-chip memory is much higher for FPGAs, and so, that is an added advantage. The power consumption of FPGA devices are much better compared to GPUs.

Unlike GPUs that offer SIMD, FPGAs can deal with irregular computations also. So, it is not necessary that all the hardware units in the FPGA have to perform the same operation. You can have different operations and different operands. This is especially useful for certain applications, where you might want to prune some calculations. For example, in your neural network, if you find that on some of the edges the weights are very small, you can prune that edge and retrain the network with the rest of the edges. There are research articles which have been able to create very heavily pruned networks, and for such networks FPGAs would have an advantage. FPGAs also support a bigger variety of data types.

The place where FPGAs are lagging behind GPUs is in terms of support for software type inputs. FPGAs have always been thought of as devices for hardware designers and so share much smaller mindshare of people who write software algorithms. However, both the major FPGA providers have seen this opportunity in machine learning and are very aggressively catching up on this aspect, in the sense that their design tools can now deal with software code as input, and are also providing integration layers for popular machine learning frameworks.

FPGAs have traditionally been used in some mission-critical applications and hence have a high degree of functional safety built-in, which is another major advantage, if you want to consider your machine learning for a critical application (such as, autonomous driving).

There have been numerous studies and articles on FPGAs vs GPU, and this book intentionally does not want to take sides. The results depend on too many factors, including whether the code is optimized for one or the other or the specific set of devices chosen for the study. Both FPGAs and GPUs are coming up with more and more features in each generation to help machine learning applications. Even for studies that seem to side with GPU, the general recommendation tends to side with the philosophy of the use of GPUs for training (a one-time activity) and FPGAs for inference (which is a repeat activity).

18.5.4 TPUs

TPUs are ASICs for *TensorFlow*-based applications – mostly for neural network type usage. These are designed and used by Google. The usage by Google is a huge endorsement by itself.

TPU architecture is different from that of GPU in that the TPUs are primarily for computation and do not have instructions for memory access themselves. In that sense, they are more like floating point instruction sets, rather than another complete parallel processing unit. For this reason, efficient data path between CPU and TPU is critical for efficiency. TPUs are 15 to 30 times faster than GPUs. Another advantage of the TPUs is that the power consumption is much smaller than CPUs as the surface area is much smaller for the processing units. For example, CPUs like Intel i5 have 20–30 CUDA cores on the die, whereas a TPU (or, GPU) can have 1000s CUDA cores (as of writing of this book – 2019).

The biggest disadvantage of ASICs for machine learning domain is the fact that machine learning algorithms are evolving, while ASICs have a fixed architecture, and don't lend themselves very easily to reprogramming at the same scale as GPUs or FPGAs. This also explains why TPUs have had a very fast evolution in terms of generation. The first generation of TPUs came in 2016, and, by 2018, we already have a third generation. From the first to the third generation, TPUs have added in high degree of parallelism, high bandwidth memory access, and support for floating point.

While Google uses TPUs, TPUs are not available for sale, and hence, this was not a hardware solution that you could consider (unless, you were a Google employee). However, recently, Google has announced access to TPUs through cloud computing. So, you could still consider using TPUs.

The other three processing solutions (traditional computers, GPUs, FPGAs) are also available through most of the popular cloud computing service providers.

Bibliography

Adhikari R, Agrawal RK (2013) An introductory study on time series modeling and forecasting. LAP Lambert Academic Publishing
Breiman L (2001) Random forests. Mach Learn J 45(1):5–32
Cortes C, Vapnik V (1995) Support-vector networks. Mach Learn 20(3):273–279
Ian G, Bengio Y, Courville A (2016) Deep learning. MIT Press, Cambridge
Karpathy A (2015) The unreasonable effectiveness of recurrent neural networks. http://karpathy.github.io/2015/05/21/rnn-effectiveness/
Kohonen T (1990) The self-organizing map. Proc IEEE 78(9):1464–1480
Moguerza JM, Muñoz A (2006) Support vector machines with applications. Stat Sci 21(3):322–336
Ng A. Machine learning. https://www.coursera.org/learn/machine-learning
Pishro-Nik H (2014) Introduction to probability, statistics, and random processes. Kappa Research, Blue Bell
Smith LI (2002) A tutorial on principal component analysis. http://www.iro.umontreal.ca/~pift6080/H09/documents/papers/pca_tutorial.pdf
Strang G (2005) Linear algebra and its applications, 4th edn. Cengage Learning
Sutton RS, Barto AG (2017) Reinforcement learning: an introduction, 2nd edn. The MIT Press, Cambridge, London
TensorFlow (n.d.) TensorFlow. https://wwwtensorfloworg/
Torch (n.d.) Torch | Scientific computing for LuaJIT. http://torchch/

G. Rebala et al., *An Introduction to Machine Learning*,
https://doi.org/10.1007/978-3-030-15729-6

Index

G. Rebala et al., *An Introduction to Machine Learning*,
https://doi.org/10.1007/978-3-030-15729-6

GPSR Compliance
The European Union's (EU) General Product Safety Regulation (GPSR) is a set of rules that requires consumer products to be safe and our obligations to ensure this.

If you have any concerns about our products, you can contact us on

ProductSafety@springernature.com

In case Publisher is established outside the EU, the EU authorized representative is:

Springer Nature Customer Service Center GmbH
Europaplatz 3
69115 Heidelberg, Germany

www.ingramcontent.com/pod-product-compliance
Lightning Source LLC
Chambersburg PA
CBHW062100160926
51535CB00037B/19
* 9 7 8 3 0 3 0 1 5 7 2 8 9 *